# Minerals in Plants

by

## Jan Scholten

D1620337

Title :        Minerals in Plants
Author:        Jan Scholten, M.D.
ISBN:          90-74817-06-8
Edition        1, March 2001
Cover design:  Peter Huurdeman
Lay-out:       Jan Scholten
Publisher:     Stichting Alonnissos
               Servaasbolwerk 13
               3512 NK, Utrecht
               The Netherlands
               Telephone   31.30.2312421
               Fax         31.30.2340211
               Email       mail@alonnissos.org
               Webpage     www.alonnissos.org
Printer:       Drukkerij Haasbeek BV
               Alphen aan den Rijn, The Netherlands

2

---

# Minerals in Plants

## Jan Scholten M.D.

*I dedicate this book to humanity*

# Acknowledgement

Most thanks go to Dolisos, Tasveld 8, Montfoort, Nederland for providing the 100 nebulisates of the plants for free. Without them this study would not have been done.
We give much thanks to Omegatech, King James Medical laboratory inc, 24700 Center Ridge Road, Suite 113, Cleveland, Ohio 44145, USA for analyzing the plant nebulisates. Especially thank go to their director Raymond J. Shamberger. Johan van Meer was so kind to look through the text and correct many mistakes and he receives many thanks for that.

# Contents

## Part 2 Elements                    **115**

# Introduction

### Purpose: comparing minerals and plants

The purpose of this book is to compare the medicinal properties of the plants and minerals. By analyzing the contents of minerals in plants we can see which minerals are comparatively high or low. Then we can compare the medicinal properties and the homeopathic pictures of a minerals high in a plant with the properties and pictures of that plant.

Or we think of a mineral for a patient but the picture isn't fitting exactly we can in the table of that mineral and see which plants are having a high content of that mineral. Or we can project the picture of a plant or aspects of it from the contents of minerals.

### Plants

The choice of the plants was made by availability. Dolisos offered nebulisate powders of 100 plants.

Analysis

The analysis of the dry plant preparations was done by Omegatech in Ohio,USA. The analysis was Induced Coupled Plasma Spectroscopy' with a Leeman® PS 1000 UVT.

### Layout

After the introducing chapters follow the plants in alphabetical order. The names used for the order were the ones used in homeopathy. Information is given about scientific, English, French and Dutch names of the plants.

Two charts give the contents of each of the 22 analyzed minerals together with their deviations. The deviations are calculated as Standard deviations from the mean of all the 100 plants. The first chart is sorted on the mineral name, the second on deviation.

After the 100 plants follow the 22 minerals. These also have 2 charts with the names of the plants, content and deviation. The first chart is sorted on plant name, the second on deviation

# Discussion

### Plant part used
The results of the analysis can depend on the part of the plant that is used for analysis. Root, leaf, flower, fruit, bark, wood can all differ in the relative content of a mineral. An example is Spirulina that has been analyzed in 2 forms and gave different results. Although the difference is not big.

### Origin of the plant
The results of the analysis can depend on the place where the plant was growing. The climate, season, stage of development the plant and soil can influence the contents. Unfortunately Dolisos wasn't able to provide this kind of information about the plants. So contamination by toxins, surrounding traffic (lead) cannot be excluded.

### Related plants
Interesting is to look at related plants and see if there are similarities in mineral content. In this study this can be done with Echinacea angustifolia and Echinacea purpurea, Tilia cordata and Tilia europea.

### Elements analyzed
In this study only 22 elements were studied. This is what Omegatech offered. The ideal would be to analyze all the elements. For example, Iodum is very much missing as an element in this study. One would expect it to be high in Spirulina and Fucus.

### Relativity of the results and deviations
The values should be seen as an indication and not absolute. This kind of analysis should be repeated often, with plants from different regions and climates, from different kinds of soil, with different varieties of plants. More minerals should be analyzed with different techniques by different laboratories

# Detection limits of the elements

Detection limits are in parts per million, ppm.

| | |
|---|---|
| Aluminium | 0.4 |
| Arsenic | 1.0 |
| Cadmium | 0.04 |
| Calcium | 1.0 |
| Chromium | 0.15 |
| Cobalt | 0.1 |
| Copper | 0.06 |
| Germanium | 0.03 |
| Iron | 0.05 |
| Lead | 0.7 |
| Lithium | 0.05 |
| Magnesium | 0.2 |
| Manganese | 0.03 |
| Mercury | 0.4 |
| Molybdenum | 0.1 |
| Nickel | 0.2 |
| Phosphorus | 2.0 |
| Potassium | 0.5 |
| Selenium | 1.0 |
| Sodium | 1.0 |

# Literature

Beauchemin D.A., Comparison of ICP atomic spectrometry techniques, Spectroscopy, Vol 7, page 12-18, 1992.

Fogg T.R., Seawater Analysis, Applied Spectroscopy, 42, 170, 1988.

Koropchak J.A., Veber M., Conver T.W., Herries J., Applied Spectroscopy, VOl 46, 1525, 1992.

Koropchak J.A., Veber M., Herries J., Spectrochima acta, Vol 47B, 825,1992.

Montaser A., Golightly D.W., Inductively Coupled Plasmas in Analytical Atomic Spectrometry, New York, 1992.

Olesik J.W., Elemental analysis using ICP-OES and ICP/MS, an evaluation and assesment of remaining problems, Analytical Chemics, Vol 63, 2A-21A, 1991.

Smith T.R., Denton M.B., Evaluation of current nebulizers and nebulizer characterization techniques, Applied Spectroscopy, 44, page 21-44, 1990.

# Part 1

# Plants

# Acorus calamus

| | |
|---|---|
| Family: | Araceae |
| Scientific name: | Acorus calamus |
| English name: | Sweet flag |
| German name: | Kalmus |
| French name: | Acore |
| Dutch name: | Kalmoes |
| Part Used: | rhizoma |
| Extraction: | watery |
| Titer: | 1/4 |
| Standard minimal: | 0,5% etheric oil |
| Comment: | high in arsenicum |

| Sorted by element name | | | Sorted by relative difference | | |
|---|---|---|---|---|---|
| aluminium | 10,6 | -0,3 | arsenicum | 8,7 | 4,0 |
| arsenicum | 8,7 | 4,0 | sodium | 6036,0 | 1,2 |
| cadmium | 0,0 | -0,3 | iron | 230,0 | 1,1 |
| calcium | 3570,0 | 0,1 | cobalt | 0,7 | 1,0 |
| chromium | 0,4 | 0,3 | nickel | 5,8 | 0,8 |
| cobalt | 0,7 | 1,0 | lithium | 1,6 | 0,5 |
| copper | 3,4 | -0,2 | phosphorus | 2213,0 | 0,5 |
| germanium | 0,0 | -0,5 | chromium | 0,4 | 0,3 |
| iron | 230,0 | 1,1 | selenium | 2,1 | 0,3 |
| lead | 0,0 | -0,4 | potassium | 29220,0 | 0,2 |
| lithium | 1,6 | 0,5 | manganese | 85,4 | 0,1 |
| magnesium | 2110,0 | 0,0 | calcium | 3570,0 | 0,1 |
| manganese | 85,4 | 0,1 | magnesium | 2110,0 | 0,0 |
| mercury | 0,0 | -0,4 | copper | 3,4 | -0,2 |
| molybdenum | 0,0 | -0,4 | cadmium | 0,0 | -0,3 |
| nickel | 5,8 | 0,8 | vanadium | 0,3 | -0,3 |
| phosphorus | 2213,0 | 0,5 | aluminium | 10,6 | -0,3 |
| potassium | 29220,0 | 0,2 | molybdenum | 0,0 | -0,4 |
| selenium | 2,1 | 0,3 | lead | 0,0 | -0,4 |
| sodium | 6036,0 | 1,2 | mercury | 0,0 | -0,4 |
| vanadium | 0,3 | -0,3 | germanium | 0,0 | -0,5 |
| zinc | 4,7 | -0,7 | zinc | 4,7 | -0,7 |

# Aesculus hippocastanum

| | |
|---|---|
| Family: | Hippocastanaceae |
| Scientific name: | Aesculus hippocastanum |
| English name: | Horse chestnut |
| German name: | Rosskastanie |
| French name: | Marronier d'inde |
| Dutch name: | Paardekastanje |
| Part Used: | semen |
| Extraction: | alcoholic watery |
| Titer: | 1/5 |
| Standard minimal: | 18% triterpeenglycosiden |
| Comment: | |

| Sorted by element name | | | Sorted by relative difference | | |
|---|---|---|---|---|---|
| aluminium | 5,7 | -0,4 | copper | 16,1 | 0,7 |
| arsenicum | 0,0 | -0,4 | phosphorus | 1544,0 | 0,0 |
| cadmium | 0,0 | -0,2 | cadmium | 0,0 | -0,2 |
| calcium | 259,1 | -0,6 | chromium | 0,1 | -0,4 |
| chromium | 0,1 | -0,4 | molybdenum | 0,0 | -0,4 |
| cobalt | 0,0 | -0,9 | aluminium | 5,7 | -0,4 |
| copper | 16,1 | 0,7 | potassium | 16470,0 | -0,4 |
| germanium | 0,0 | -0,5 | selenium | 0,0 | -0,4 |
| iron | 2,6 | -0,5 | manganese | 1,6 | -0,4 |
| lead | 0,0 | -0,4 | nickel | 0,8 | -0,4 |
| lithium | 0,0 | -0,5 | arsenicum | 0,0 | -0,4 |
| magnesium | 824,2 | -0,7 | lead | 0,0 | -0,4 |
| manganese | 1,6 | -0,4 | mercury | 0,0 | -0,4 |
| mercury | 0,0 | -0,4 | sodium | 491,5 | -0,5 |
| molybdenum | 0,0 | -0,4 | iron | 2,6 | -0,5 |
| nickel | 0,8 | -0,4 | lithium | 0,0 | -0,5 |
| phosphorus | 1544,0 | 0,0 | germanium | 0,0 | -0,5 |
| potassium | 16470,0 | -0,4 | vanadium | 0,1 | -0,5 |
| selenium | 0,0 | -0,4 | calcium | 259,1 | -0,6 |
| sodium | 491,5 | -0,5 | zinc | 5,8 | -0,6 |
| vanadium | 0,1 | -0,5 | magnesium | 824,2 | -0,7 |
| zinc | 5,8 | -0,6 | cobalt | 0,0 | -0,9 |

# Agnus castus

| | |
|---|---|
| Family: | Verbenaceae |
| Scientific name: | Vitex agnus castus |
| English name: | Chaste tree |
| German name: | Keuschstrauch |
| French name: | Gattilier |
| Dutch name: | Monnikspeper |
| Part Used: | semen |
| Extraction: | watery |
| Titer: | 1/10 |
| Standard minimal: | |
| Comment: | remarkable is the general lack of |

minerals. This fits the general lack of stamina in the remedy.

| Sorted by element name | | | Sorted by relative difference | | |
|---|---|---|---|---|---|
| aluminium | 28,0 | 0,0 | vanadium | 0,5 | 0,0 |
| arsenicum | 0,7 | -0,1 | aluminium | 28,0 | 0,0 |
| cadmium | 0,0 | -0,3 | arsenicum | 0,7 | -0,1 |
| calcium | 2418,0 | -0,2 | chromium | 0,2 | -0,2 |
| chromium | 0,2 | -0,2 | calcium | 2418,0 | -0,2 |
| cobalt | 0,2 | -0,3 | iron | 36,4 | -0,2 |
| copper | 3,0 | -0,2 | phosphorus | 1216,0 | -0,2 |
| germanium | 0,0 | -0,5 | potassium | 19780,0 | -0,2 |
| iron | 36,4 | -0,2 | copper | 3,0 | -0,2 |
| lead | 0,0 | -0,4 | cadmium | 0,0 | -0,3 |
| lithium | 0,0 | -0,5 | cobalt | 0,2 | -0,3 |
| magnesium | 1098,0 | -0,6 | molybdenum | 0,0 | -0,4 |
| manganese | 5,5 | -0,4 | sodium | 788,4 | -0,4 |
| mercury | 0,0 | -0,4 | selenium | 0,0 | -0,4 |
| molybdenum | 0,0 | -0,4 | manganese | 5,5 | -0,4 |
| nickel | 0,1 | -0,6 | lead | 0,0 | -0,4 |
| phosphorus | 1216,0 | -0,2 | mercury | 0,0 | -0,4 |
| potassium | 19780,0 | -0,2 | lithium | 0,0 | -0,5 |
| selenium | 0,0 | -0,4 | germanium | 0,0 | -0,5 |
| sodium | 788,4 | -0,4 | magnesium | 1098,0 | -0,6 |
| vanadium | 0,5 | 0,0 | nickel | 0,1 | -0,6 |
| zinc | 5,3 | -0,6 | zinc | 5,3 | -0,6 |

# Alchemilla vulgaris

Family:                      Rosaceae
Scientific name:             Alchemilla vulgaris
English name:                Ladys mantle
German name:                 Gemeinsamer Frauenmantel
French name:                 Manteau de Notre-Dame
Dutch name:                  Vrouwenmantel
Part Used:                   herba
Extraction:                  alcoholic watery
Titer:                       1/5
Standard minimal:
Comment:

| Sorted by element name | | | Sorted by relative difference | | |
|---|---|---|---|---|---|
| aluminium | 4,7 | -0,4 | germanium | 1,0 | 0,8 |
| arsenicum | 0,0 | -0,4 | zinc | 29,2 | 0,7 |
| cadmium | 0,0 | -0,3 | magnesium | 2700,0 | 0,3 |
| calcium | 2562,0 | -0,1 | nickel | 3,3 | 0,2 |
| chromium | 0,1 | -0,4 | cobalt | 0,4 | 0,2 |
| cobalt | 0,4 | 0,2 | phosphorus | 1824,0 | 0,2 |
| copper | 6,3 | 0,0 | copper | 6,3 | 0,0 |
| germanium | 1,0 | 0,8 | lithium | 0,7 | 0,0 |
| iron | 15,4 | -0,4 | manganese | 58,1 | 0,0 |
| lead | 0,0 | -0,4 | potassium | 23200,0 | -0,1 |
| lithium | 0,7 | 0,0 | calcium | 2562,0 | -0,1 |
| magnesium | 2700,0 | 0,3 | cadmium | 0,0 | -0,3 |
| manganese | 58,1 | 0,0 | chromium | 0,1 | -0,4 |
| mercury | 0,0 | -0,4 | molybdenum | 0,0 | -0,4 |
| molybdenum | 0,0 | -0,4 | iron | 15,4 | -0,4 |
| nickel | 3,3 | 0,2 | aluminium | 4,7 | -0,4 |
| phosphorus | 1824,0 | 0,2 | selenium | 0,0 | -0,4 |
| potassium | 23200,0 | -0,1 | arsenicum | 0,0 | -0,4 |
| selenium | 0,0 | -0,4 | lead | 0,0 | -0,4 |
| sodium | 94,0 | -0,6 | mercury | 0,0 | -0,4 |
| vanadium | 0,0 | -0,7 | sodium | 94,0 | -0,6 |
| zinc | 29,2 | 0,7 | vanadium | 0,0 | -0,7 |

# Allium sativum

| | |
|---|---|
| Family: | Liliaceae |
| Scientific name: | Allium sativum |
| English name: | Garlic |
| German name: | Knoblauch |
| French name: | Ail |
| Dutch name: | Knoflook |
| Part Used: | bulbus |
| Extraction: | watery |
| Titer: | 1/5 |
| Standard minimal: | |
| Comment: | high in phosphorus |

| Sorted by element name | | | Sorted by relative difference | | |
|---|---|---|---|---|---|
| aluminium | 6,2 | -0,4 | phosphorus | 4412,0 | 2,0 |
| arsenicum | 1,4 | 0,3 | zinc | 22,0 | 0,3 |
| cadmium | 0,0 | -0,3 | arsenicum | 1,4 | 0,3 |
| calcium | 1021,0 | -0,5 | copper | 5,7 | 0,0 |
| chromium | 0,1 | -0,4 | lithium | 0,4 | -0,2 |
| cobalt | 0,0 | -0,9 | sodium | 1198,0 | -0,2 |
| copper | 5,7 | 0,0 | cadmium | 0,0 | -0,3 |
| germanium | 0,0 | -0,5 | potassium | 17310,0 | -0,3 |
| iron | 14,5 | -0,4 | manganese | 12,2 | -0,4 |
| lead | 0,0 | -0,4 | molybdenum | 0,0 | -0,4 |
| lithium | 0,4 | -0,2 | aluminium | 6,2 | -0,4 |
| magnesium | 953,1 | -0,6 | iron | 14,5 | -0,4 |
| manganese | 12,2 | -0,4 | selenium | 0,0 | -0,4 |
| mercury | 0,0 | -0,4 | chromium | 0,1 | -0,4 |
| molybdenum | 0,0 | -0,4 | lead | 0,0 | -0,4 |
| nickel | 0,0 | -0,6 | mercury | 0,0 | -0,4 |
| phosphorus | 4412,0 | 2,0 | calcium | 1021,0 | -0,5 |
| potassium | 17310,0 | -0,3 | germanium | 0,0 | -0,5 |
| selenium | 0,0 | -0,4 | magnesium | 953,1 | -0,6 |
| sodium | 1198,0 | -0,2 | nickel | 0,0 | -0,6 |
| vanadium | 0,0 | -0,7 | vanadium | 0,0 | -0,7 |
| zinc | 22,0 | 0,3 | cobalt | 0,0 | -0,9 |

# Althea officinalis

| | |
|---|---|
| Family: | Malvaceae |
| Scientific name: | Althea officinalis |
| English name: | Marsh mallow |
| German name: | Eibisch |
| French name: | Guimauve |
| Dutch name: | Heemst |
| Part Used: | radix |
| Extraction: | watery |
| Titer: | 1/4 |
| Standard minimal: | |
| Comment: | |

| Sorted by element name | | | Sorted by relative difference | | |
|---|---|---|---|---|---|
| aluminium | 0,7 | -0,5 | molybdenum | 2,2 | 2,8 |
| arsenicum | 0,0 | -0,4 | vanadium | 1,3 | 1,2 |
| cadmium | 0,0 | -0,3 | cobalt | 0,5 | 0,4 |
| calcium | 368,2 | -0,6 | germanium | 0,4 | 0,0 |
| chromium | 0,2 | -0,2 | magnesium | 2202,0 | 0,0 |
| cobalt | 0,5 | 0,4 | chromium | 0,2 | -0,2 |
| copper | 0,0 | -0,5 | lithium | 0,4 | -0,2 |
| germanium | 0,4 | 0,0 | cadmium | 0,0 | -0,3 |
| iron | 20,6 | -0,3 | phosphorus | 1110,0 | -0,3 |
| lead | 0,0 | -0,4 | iron | 20,6 | -0,3 |
| lithium | 0,4 | -0,2 | manganese | 7,4 | -0,4 |
| magnesium | 2202,0 | 0,0 | selenium | 0,0 | -0,4 |
| manganese | 7,4 | -0,4 | nickel | 0,9 | -0,4 |
| mercury | 0,0 | -0,4 | arsenicum | 0,0 | -0,4 |
| molybdenum | 2,2 | 2,8 | lead | 0,0 | -0,4 |
| nickel | 0,9 | -0,4 | mercury | 0,0 | -0,4 |
| phosphorus | 1110,0 | -0,3 | aluminium | 0,7 | -0,5 |
| potassium | 80,7 | -1,1 | copper | 0,0 | -0,5 |
| selenium | 0,0 | -0,4 | sodium | 155,1 | -0,6 |
| sodium | 155,1 | -0,6 | zinc | 6,8 | -0,6 |
| vanadium | 1,3 | 1,2 | calcium | 368,2 | -0,6 |
| zinc | 6,8 | -0,6 | potassium | 80,7 | -1,1 |

# Ananassa comosus

| | |
|---|---|
| Family: | Bromeliaceae |
| Scientific name: | Ananassa Comosus |
| English name: | Pine-apple |
| German name: | Ananas |
| French name: | Ananas |
| Dutch name: | Ananas |
| Part Used: | fructus |
| Extraction: | watery |
| Titer: | 1/5 |
| Standard minimal: | 4,5% bromelaine 80 enzymatic units |
| Comment: | |

| Sorted by element name | | | Sorted by relative difference | | |
|---|---|---|---|---|---|
| aluminium | 5,0 | -0,4 | arsenicum | 3,1 | 1,1 |
| arsenicum | 3,1 | 1,1 | vanadium | 0,8 | 0,5 |
| cadmium | 0,0 | -0,3 | calcium | 4730,0 | 0,3 |
| calcium | 4730,0 | 0,3 | manganese | 81,5 | 0,1 |
| chromium | 0,0 | -0,6 | copper | 4,5 | -0,1 |
| cobalt | 0,0 | -0,9 | nickel | 1,9 | -0,2 |
| copper | 4,5 | -0,1 | phosphorus | 1250,0 | -0,2 |
| germanium | 0,0 | -0,5 | cadmium | 0,0 | -0,3 |
| iron | 12,2 | -0,4 | molybdenum | 0,0 | -0,4 |
| lead | 0,0 | -0,4 | aluminium | 5,0 | -0,4 |
| lithium | 0,0 | -0,5 | iron | 12,2 | -0,4 |
| magnesium | 601,0 | -0,8 | selenium | 0,0 | -0,4 |
| manganese | 81,5 | 0,1 | lead | 0,0 | -0,4 |
| mercury | 0,0 | -0,4 | mercury | 0,0 | -0,4 |
| molybdenum | 0,0 | -0,4 | lithium | 0,0 | -0,5 |
| nickel | 1,9 | -0,2 | germanium | 0,0 | -0,5 |
| phosphorus | 1250,0 | -0,2 | zinc | 7,4 | -0,5 |
| potassium | 10060,0 | -0,6 | chromium | 0,0 | -0,6 |
| selenium | 0,0 | -0,4 | sodium | 16,3 | -0,6 |
| sodium | 16,3 | -0,6 | potassium | 10060,0 | -0,6 |
| vanadium | 0,8 | 0,5 | magnesium | 601,0 | -0,8 |
| zinc | 7,4 | -0,5 | cobalt | 0,0 | -0,9 |

# Angelica archangelica

| | |
|---|---|
| Family: | Umbelliferae |
| Scientific name: | Angelica archangelica |
| English name: | Holy ghost, Angelica root |
| German name: | Engelwurz |
| French name: | Herbe du Saint Esprit |
| Dutch name: | Engelwortel |
| Part Used: | radix |
| Extraction: | alcoholic watery |
| Titer: | 1/4 |
| Standard minimal: | |
| Comment: | |

| Sorted by element name | | | Sorted by relative difference | | |
|---|---|---|---|---|---|
| aluminium | 18,4 | -0,2 | mercury | 2,0 | 3,6 |
| arsenicum | 0,0 | -0,4 | lead | 0,8 | 0,4 |
| cadmium | 0,1 | -0,1 | germanium | 0,3 | -0,1 |
| calcium | 284,0 | -0,6 | cadmium | 0,1 | -0,1 |
| chromium | 0,1 | -0,5 | aluminium | 18,4 | -0,2 |
| cobalt | 0,2 | -0,3 | phosphorus | 1231,0 | -0,2 |
| copper | 1,2 | -0,4 | iron | 27,1 | -0,3 |
| germanium | 0,3 | -0,1 | cobalt | 0,2 | -0,3 |
| iron | 27,1 | -0,3 | molybdenum | 0,0 | -0,4 |
| lead | 0,8 | 0,4 | copper | 1,2 | -0,4 |
| lithium | 0,0 | -0,5 | selenium | 0,0 | -0,4 |
| magnesium | 491,6 | -0,9 | manganese | 4,7 | -0,4 |
| manganese | 4,7 | -0,4 | arsenicum | 0,0 | -0,4 |
| mercury | 2,0 | 3,6 | potassium | 14530,0 | -0,5 |
| molybdenum | 0,0 | -0,4 | lithium | 0,0 | -0,5 |
| nickel | 0,3 | -0,6 | chromium | 0,1 | -0,5 |
| phosphorus | 1231,0 | -0,2 | sodium | 260,1 | -0,5 |
| potassium | 14530,0 | -0,5 | nickel | 0,3 | -0,6 |
| selenium | 0,0 | -0,4 | calcium | 284,0 | -0,6 |
| sodium | 260,1 | -0,5 | vanadium | 0,0 | -0,7 |
| vanadium | 0,0 | -0,7 | zinc | 3,5 | -0,7 |
| zinc | 3,5 | -0,7 | magnesium | 491,6 | -0,9 |

# Arctium lappa

Family:               Compositae
Scientific name:      Arctium lappa
English name:         Burdock
German name:          Filzige Klette
French name:          Gouteron
Dutch name:           Klis grote
Part Used:            radix
Extraction:           watery
Titer:                1/4
Standard minimal:
Comment:              The picture of Arctium lappa has
much in common with Chromium. According to this analysis
nickel and coblat are higher.

| Sorted by element name | | | Sorted by relative difference | | |
|---|---|---|---|---|---|
| aluminium | 26,4 | -0,1 | nickel | 11,5 | 2,3 |
| arsenicum | 2,7 | 0,9 | cobalt | 1,0 | 1,7 |
| cadmium | 0,0 | -0,3 | arsenicum | 2,7 | 0,9 |
| calcium | 1319,0 | -0,4 | chromium | 0,7 | 0,9 |
| chromium | 0,7 | 0,9 | lead | 0,7 | 0,3 |
| cobalt | 1,0 | 1,7 | iron | 71,1 | 0,0 |
| copper | 4,4 | -0,1 | potassium | 24720,0 | 0,0 |
| germanium | 0,0 | -0,5 | phosphorus | 1477,0 | -0,1 |
| iron | 71,1 | 0,0 | aluminium | 26,4 | -0,1 |
| lead | 0,7 | 0,3 | copper | 4,4 | -0,1 |
| lithium | 0,5 | -0,2 | lithium | 0,5 | -0,2 |
| magnesium | 1342,0 | -0,4 | sodium | 1511,0 | -0,2 |
| manganese | 20,5 | -0,3 | cadmium | 0,0 | -0,3 |
| mercury | 0,0 | -0,4 | manganese | 20,5 | -0,3 |
| molybdenum | 0,0 | -0,4 | molybdenum | 0,0 | -0,4 |
| nickel | 11,5 | 2,3 | calcium | 1319,0 | -0,4 |
| phosphorus | 1477,0 | -0,1 | selenium | 0,0 | -0,4 |
| potassium | 24720,0 | 0,0 | magnesium | 1342,0 | -0,4 |
| selenium | 0,0 | -0,4 | mercury | 0,0 | -0,4 |
| sodium | 1511,0 | -0,2 | germanium | 0,0 | -0,5 |
| vanadium | 0,0 | -0,7 | zinc | 6,5 | -0,6 |
| zinc | 6,5 | -0,6 | vanadium | 0,0 | -0,7 |

# Ballota foetida

Family:              Labiatae
Scientific name:     Ballota foetida
English name:        Black horehound
German name:         Schwarznessel
French name:         Ballote noire
Dutch name:          Stinkend ballota
Part Used:           herba
Extraction:          watery
Titer:               1/6
Standard minimal:    spores betaine
Comment:

| Sorted by element name | | | Sorted by relative difference | | |
|---|---|---|---|---|---|
| aluminium | 18,5 | -0,2 | molybdenum | 1,3 | 1,5 |
| arsenicum | 1,8 | 0,5 | potassium | 42860,0 | 0,7 |
| cadmium | 0,1 | -0,1 | arsenicum | 1,8 | 0,5 |
| calcium | 1753,0 | -0,3 | vanadium | 0,8 | 0,5 |
| chromium | 0,3 | 0,1 | magnesium | 3065,0 | 0,5 |
| cobalt | 0,3 | -0,1 | phosphorus | 1790,0 | 0,2 |
| copper | 4,7 | -0,1 | lithium | 1,0 | 0,2 |
| germanium | 0,0 | -0,5 | iron | 82,0 | 0,1 |
| iron | 82,0 | 0,1 | chromium | 0,3 | 0,1 |
| lead | 0,5 | 0,1 | lead | 0,5 | 0,1 |
| lithium | 1,0 | 0,2 | zinc | 17,1 | 0,0 |
| magnesium | 3065,0 | 0,5 | cadmium | 0,1 | -0,1 |
| manganese | 51,1 | -0,1 | cobalt | 0,3 | -0,1 |
| mercury | 0,0 | -0,4 | manganese | 51,1 | -0,1 |
| molybdenum | 1,3 | 1,5 | copper | 4,7 | -0,1 |
| nickel | 0,0 | -0,6 | aluminium | 18,5 | -0,2 |
| phosphorus | 1790,0 | 0,2 | calcium | 1753,0 | -0,3 |
| potassium | 42860,0 | 0,7 | sodium | 837,0 | -0,4 |
| selenium | 0,0 | -0,4 | selenium | 0,0 | -0,4 |
| sodium | 837,0 | -0,4 | mercury | 0,0 | -0,4 |
| vanadium | 0,8 | 0,5 | germanium | 0,0 | -0,5 |
| zinc | 17,1 | 0,0 | nickel | 0,0 | -0,6 |

# Berberis vulgaris

| | |
|---|---|
| Family: | Berberidaceae |
| Scientific name: | Berberis vulgaris |
| English name: | Barberry |
| German name: | Sauerdorn, Berberitze |
| French name: | Epine vinette |
| Dutch name: | Zuurbes |
| Part Used: | cortex radicis |
| Extraction: | alcoholic watery |
| Titer: | 1/6 |
| Standard minimal: | 1% berberine |
| Comment: | |

| Sorted by element name | | | Sorted by relative difference | | |
|---|---|---|---|---|---|
| aluminium | 6,9 | -0,4 | cobalt | 0,9 | 1,5 |
| arsenicum | 0,0 | -0,4 | germanium | 1,4 | 1,3 |
| cadmium | 0,1 | 0,1 | molybdenum | 0,8 | 0,8 |
| calcium | 1427,0 | -0,4 | selenium | 2,4 | 0,4 |
| chromium | 0,2 | -0,2 | zinc | 21,9 | 0,3 |
| cobalt | 0,9 | 1,5 | vanadium | 0,6 | 0,2 |
| copper | 6,2 | 0,0 | cadmium | 0,1 | 0,1 |
| germanium | 1,4 | 1,3 | copper | 6,2 | 0,0 |
| iron | 17,8 | -0,4 | sodium | 1723,0 | -0,1 |
| lead | 0,0 | -0,4 | chromium | 0,2 | -0,2 |
| lithium | 0,5 | -0,2 | lithium | 0,5 | -0,2 |
| magnesium | 1602,0 | -0,3 | manganese | 28,2 | -0,2 |
| manganese | 28,2 | -0,2 | magnesium | 1602,0 | -0,3 |
| mercury | 0,0 | -0,4 | iron | 17,8 | -0,4 |
| molybdenum | 0,8 | 0,8 | aluminium | 6,9 | -0,4 |
| nickel | 0,0 | -0,6 | calcium | 1427,0 | -0,4 |
| phosphorus | 731,7 | -0,6 | arsenicum | 0,0 | -0,4 |
| potassium | 12510,0 | -0,5 | lead | 0,0 | -0,4 |
| selenium | 2,4 | 0,4 | mercury | 0,0 | -0,4 |
| sodium | 1723,0 | -0,1 | potassium | 12510,0 | -0,5 |
| vanadium | 0,6 | 0,2 | phosphorus | 731,7 | -0,6 |
| zinc | 21,9 | 0,3 | nickel | 0,0 | -0,6 |

# Beta vulgaris

Family:              Chenopodiaceae
Scientific name:     Beta vulgaris
English name:        Common beet
German name:         Runkelrübe
French name:         Bette suavage
Dutch name:          Biet
Part Used:           succus radicis
Extraction:          watery
Titer:               1/20
Standard minimal:    sporen betaine
Comment:

| Sorted by element name | | | Sorted by relative difference | | |
|---|---|---|---|---|---|
| aluminium | 5,5 | -0,4 | selenium | 9,1 | 2,4 |
| arsenicum | 0,0 | -0,4 | zinc | 27,7 | 0,6 |
| cadmium | 0,0 | -0,1 | mercury | 0,5 | 0,6 |
| calcium | 4362,0 | 0,2 | phosphorus | 2124,0 | 0,4 |
| chromium | 0,0 | -0,5 | calcium | 4362,0 | 0,2 |
| cobalt | 0,1 | -0,6 | lithium | 1,1 | 0,2 |
| copper | 5,3 | -0,1 | sodium | 2049,0 | 0,0 |
| germanium | 0,0 | -0,5 | copper | 5,3 | -0,1 |
| iron | 59,3 | -0,1 | iron | 59,3 | -0,1 |
| lead | 0,0 | -0,4 | cadmium | 0,0 | -0,1 |
| lithium | 1,1 | 0,2 | manganese | 28,6 | -0,2 |
| magnesium | 1164,0 | -0,5 | potassium | 19170,0 | -0,3 |
| manganese | 28,6 | -0,2 | nickel | 1,5 | -0,3 |
| mercury | 0,5 | 0,6 | molybdenum | 0,0 | -0,4 |
| molybdenum | 0,0 | -0,4 | aluminium | 5,5 | -0,4 |
| nickel | 1,5 | -0,3 | arsenicum | 0,0 | -0,4 |
| phosphorus | 2124,0 | 0,4 | lead | 0,0 | -0,4 |
| potassium | 19170,0 | -0,3 | germanium | 0,0 | -0,5 |
| selenium | 9,1 | 2,4 | chromium | 0,0 | -0,5 |
| sodium | 2049,0 | 0,0 | magnesium | 1164,0 | -0,5 |
| vanadium | 0,1 | -0,6 | vanadium | 0,1 | -0,6 |
| zinc | 27,7 | 0,6 | cobalt | 0,1 | -0,6 |

# Betula alba

| | |
|---|---|
| Family: | Betulaceae |
| Scientific name: | Betula alba |
| English name: | White birch |
| German name: | Weissbirke |
| French name: | Bouleau blanc |
| Dutch name: | Berk witte |
| Part Used: | folia |
| Extraction: | watery |
| Titer: | 1/4,5 |
| Standard minimal: | 1% hyperoside |
| Comment: | |

| Sorted by element name | | | Sorted by relative difference | | |
|---|---|---|---|---|---|
| aluminium | 15,8 | -0,2 | manganese | 712,6 | 4,4 |
| arsenicum | 0,0 | -0,4 | nickel | 18,0 | 3,9 |
| cadmium | 0,0 | -0,3 | selenium | 8,2 | 2,1 |
| calcium | 2855,0 | -0,1 | zinc | 51,4 | 2,0 |
| chromium | 0,4 | 0,3 | magnesium | 5195,0 | 1,5 |
| cobalt | 0,5 | 0,4 | phosphorus | 3052,0 | 1,0 |
| copper | 1,2 | -0,4 | lead | 1,0 | 0,6 |
| germanium | 0,0 | -0,5 | cobalt | 0,5 | 0,4 |
| iron | 21,8 | -0,3 | chromium | 0,4 | 0,3 |
| lead | 1,0 | 0,6 | lithium | 1,1 | 0,2 |
| lithium | 1,1 | 0,2 | potassium | 24910,0 | 0,0 |
| magnesium | 5195,0 | 1,5 | calcium | 2855,0 | -0,1 |
| manganese | 712,6 | 4,4 | aluminium | 15,8 | -0,2 |
| mercury | 0,0 | -0,4 | sodium | 1226,0 | -0,2 |
| molybdenum | 0,0 | -0,4 | cadmium | 0,0 | -0,3 |
| nickel | 18,0 | 3,9 | iron | 21,8 | -0,3 |
| phosphorus | 3052,0 | 1,0 | molybdenum | 0,0 | -0,4 |
| potassium | 24910,0 | 0,0 | copper | 1,2 | -0,4 |
| selenium | 8,2 | 2,1 | arsenicum | 0,0 | -0,4 |
| sodium | 1226,0 | -0,2 | mercury | 0,0 | -0,4 |
| vanadium | 0,0 | -0,7 | germanium | 0,0 | -0,5 |
| zinc | 51,4 | 2,0 | vanadium | 0,0 | -0,7 |

# Boldo fragrans

| | |
|---|---|
| Family: | Monimiaceae |
| Scientific name: | Boldo |
| English name: | Boldo |
| German name: | Boldobaum |
| French name: | Boldo, Laurel de Chile |
| Dutch name: | Boldo |
| Part Used: | folia |
| Extraction: | alcoholic watery |
| Titer: | 1/3 |
| Standard minimal: | 0,1% boldin |
| Comment: | |

| Sorted by element name | | | Sorted by relative difference | | |
|---|---|---|---|---|---|
| aluminium | 4,2 | -0,4 | molybdenum | 1,0 | 1,1 |
| arsenicum | 0,0 | -0,4 | germanium | 1,2 | 1,1 |
| cadmium | 0,0 | -0,3 | selenium | 3,5 | 0,7 |
| calcium | 402,9 | -0,6 | cadmium | 0,0 | -0,3 |
| chromium | 0,1 | -0,5 | mercury | 0,1 | -0,3 |
| cobalt | 0,0 | -0,9 | lithium | 0,2 | -0,3 |
| copper | 0,5 | -0,4 | iron | 14,5 | -0,4 |
| germanium | 1,2 | 1,1 | manganese | 6,7 | -0,4 |
| iron | 14,5 | -0,4 | aluminium | 4,2 | -0,4 |
| lead | 0,0 | -0,4 | copper | 0,5 | -0,4 |
| lithium | 0,2 | -0,3 | arsenicum | 0,0 | -0,4 |
| magnesium | 456,5 | -0,9 | lead | 0,0 | -0,4 |
| manganese | 6,7 | -0,4 | chromium | 0,1 | -0,5 |
| mercury | 0,1 | -0,3 | phosphorus | 843,6 | -0,5 |
| molybdenum | 1,0 | 1,1 | sodium | 163,5 | -0,6 |
| nickel | 0,0 | -0,6 | calcium | 402,9 | -0,6 |
| phosphorus | 843,6 | -0,5 | nickel | 0,0 | -0,6 |
| potassium | 7821,0 | -0,7 | vanadium | 0,0 | -0,7 |
| selenium | 3,5 | 0,7 | potassium | 7821,0 | -0,7 |
| sodium | 163,5 | -0,6 | zinc | 2,4 | -0,8 |
| vanadium | 0,0 | -0,7 | cobalt | 0,0 | -0,9 |
| zinc | 2,4 | -0,8 | magnesium | 456,5 | -0,9 |

# Carduus marianus

| | |
|---|---|
| Family: | Compositae |
| Scientific name: | Carduus marianus |
| English name: | St.Mary's thistle |
| German name: | Mariendistel, Gallendistel |
| French name: | CHardon marie |
| Dutch name: | Mariadistel |
| Part Used: | fructus |
| Extraction: | alcoholic watery |
| Titer: | 1/10 |
| Standard minimal: | 10% silymarine |
| Comment: | the very high Cadmium content is |

remarkable. In common is the feeling of confidence and
aversion to be helped.

| Sorted by element name | | | Sorted by relative difference | | |
|---|---|---|---|---|---|
| aluminium | 1,9 | -0,4 | cadmium | 3,0 | 9,4 |
| arsenicum | 1,9 | 0,5 | arsenicum | 1,9 | 0,5 |
| cadmium | 3,0 | 9,4 | selenium | 2,3 | 0,3 |
| calcium | 191,3 | -0,6 | germanium | 0,5 | 0,2 |
| chromium | 0,2 | -0,2 | chromium | 0,2 | -0,2 |
| cobalt | 0,0 | -0,9 | lithium | 0,3 | -0,3 |
| copper | 1,9 | -0,3 | sodium | 1072,0 | -0,3 |
| germanium | 0,5 | 0,2 | copper | 1,9 | -0,3 |
| iron | 21,5 | -0,3 | iron | 21,5 | -0,3 |
| lead | 0,0 | -0,4 | molybdenum | 0,0 | -0,4 |
| lithium | 0,3 | -0,3 | nickel | 0,9 | -0,4 |
| magnesium | 22,5 | -1,1 | aluminium | 1,9 | -0,4 |
| manganese | 0,2 | -0,4 | manganese | 0,2 | -0,4 |
| mercury | 0,0 | -0,4 | lead | 0,0 | -0,4 |
| molybdenum | 0,0 | -0,4 | mercury | 0,0 | -0,4 |
| nickel | 0,9 | -0,4 | calcium | 191,3 | -0,6 |
| phosphorus | 73,7 | -1,0 | vanadium | 0,0 | -0,7 |
| potassium | 256,7 | -1,0 | zinc | 3,8 | -0,7 |
| selenium | 2,3 | 0,3 | cobalt | 0,0 | -0,9 |
| sodium | 1072,0 | -0,3 | phosphorus | 73,7 | -1,0 |
| vanadium | 0,0 | -0,7 | potassium | 256,7 | -1,0 |
| zinc | 3,8 | -0,7 | magnesium | 22,5 | -1,1 |

# Carica papaya

| | |
|---|---|
| Family: | Caricaceae |
| Scientific name: | Carica papaya |
| English name: | Common papaw |
| German name: | Melonenbaum |
| French name: | Papayer |
| Dutch name: | Papaya |
| Part Used: | fructus |
| Extraction: | watery |
| Titer: | 1/2,5 |
| Standard minimal: | |
| Comment: | |

| Sorted by element name | | | Sorted by relative difference | | |
|---|---|---|---|---|---|
| aluminium | 4,3 | -0,4 | lead | 4,5 | 4,1 |
| arsenicum | 3,3 | 1,2 | mercury | 1,6 | 2,8 |
| cadmium | 0,0 | -0,3 | arsenicum | 3,3 | 1,2 |
| calcium | 3270,0 | 0,0 | magnesium | 2320,0 | 0,1 |
| chromium | 0,1 | -0,5 | vanadium | 0,5 | 0,0 |
| cobalt | 0,0 | -0,9 | calcium | 3270,0 | 0,0 |
| copper | 1,1 | -0,4 | sodium | 1843,0 | -0,1 |
| germanium | 0,0 | -0,5 | potassium | 21620,0 | -0,2 |
| iron | 10,9 | -0,4 | phosphorus | 1314,0 | -0,2 |
| lead | 4,5 | 4,1 | cadmium | 0,0 | -0,3 |
| lithium | 0,3 | -0,3 | lithium | 0,3 | -0,3 |
| magnesium | 2320,0 | 0,1 | molybdenum | 0,0 | -0,4 |
| manganese | 9,4 | -0,4 | manganese | 9,4 | -0,4 |
| mercury | 1,6 | 2,8 | copper | 1,1 | -0,4 |
| molybdenum | 0,0 | -0,4 | aluminium | 4,3 | -0,4 |
| nickel | 0,2 | -0,6 | selenium | 0,0 | -0,4 |
| phosphorus | 1314,0 | -0,2 | iron | 10,9 | -0,4 |
| potassium | 21620,0 | -0,2 | chromium | 0,1 | -0,5 |
| selenium | 0,0 | -0,4 | germanium | 0,0 | -0,5 |
| sodium | 1843,0 | -0,1 | nickel | 0,2 | -0,6 |
| vanadium | 0,5 | 0,0 | zinc | 3,8 | -0,7 |
| zinc | 3,8 | -0,7 | cobalt | 0,0 | -0,9 |

# Caroube

Family:
Scientific name:        Caroube
English name:
German name:
French name:        Caroube
Dutch name:        Johannesbroodboom
Part Used:
Extraction:
Titer:        1/
Standard minimal:
Comment:

| Sorted by element name | | | Sorted by relative difference | | |
|---|---|---|---|---|---|
| aluminium | 450,8 | 6,4 | iron | 1178,0 | 7,6 |
| arsenicum | 0,0 | -0,4 | aluminium | 450,8 | 6,4 |
| cadmium | 0,0 | -0,3 | chromium | 1,9 | 3,5 |
| calcium | 3502,0 | 0,1 | vanadium | 2,8 | 3,4 |
| chromium | 1,9 | 3,5 | cobalt | 1,1 | 2,0 |
| cobalt | 1,1 | 2,0 | mercury | 0,7 | 1,0 |
| copper | 5,6 | -0,1 | nickel | 4,2 | 0,4 |
| germanium | 0,0 | -0,5 | calcium | 3502,0 | 0,1 |
| iron | 1178,0 | 7,6 | copper | 5,6 | -0,1 |
| lead | 0,0 | -0,4 | manganese | 27,2 | -0,3 |
| lithium | 0,0 | -0,5 | cadmium | 0,0 | -0,3 |
| magnesium | 797,0 | -0,7 | molybdenum | 0,0 | -0,4 |
| manganese | 27,2 | -0,3 | selenium | 0,0 | -0,4 |
| mercury | 0,7 | 1,0 | arsenicum | 0,0 | -0,4 |
| molybdenum | 0,0 | -0,4 | lead | 0,0 | -0,4 |
| nickel | 4,2 | 0,4 | lithium | 0,0 | -0,5 |
| phosphorus | 438,4 | -0,8 | zinc | 8,4 | -0,5 |
| potassium | 7752,0 | -0,7 | germanium | 0,0 | -0,5 |
| selenium | 0,0 | -0,4 | sodium | 91,8 | -0,6 |
| sodium | 91,8 | -0,6 | magnesium | 797,0 | -0,7 |
| vanadium | 2,8 | 3,4 | potassium | 7752,0 | -0,7 |
| zinc | 8,4 | -0,5 | phosphorus | 438,4 | -0,8 |

# Carragheen

| | |
|---|---|
| Family: | Gigartinaceae |
| Scientific name: | Chondrus crispus |
| English name: | Carragheen |
| German name: | Karrageen |
| French name: | Mousse d'Irlande |
| Dutch name: | Iers mos |
| Part Used: | thallus |
| Extraction: | |
| Titer: | 1/ |
| Standard minimal: | |
| Comment: | |

| Sorted by element name | | | Sorted by relative difference | | |
|---|---|---|---|---|---|
| aluminium | 20,1 | -0,2 | sodium | 21350,0 | 5,7 |
| arsenicum | 0,0 | -0,4 | germanium | 4,4 | 5,2 |
| cadmium | 0,3 | 0,7 | zinc | 46,4 | 1,7 |
| calcium | 3302,0 | 0,0 | vanadium | 1,3 | 1,2 |
| chromium | 0,0 | -0,6 | cobalt | 0,7 | 1,0 |
| cobalt | 0,7 | 1,0 | cadmium | 0,3 | 0,7 |
| copper | 1,1 | -0,4 | iron | 168,6 | 0,7 |
| germanium | 4,4 | 5,2 | magnesium | 3327,0 | 0,6 |
| iron | 168,6 | 0,7 | nickel | 3,2 | 0,2 |
| lead | 0,0 | -0,4 | calcium | 3302,0 | 0,0 |
| lithium | 0,0 | -0,5 | aluminium | 20,1 | -0,2 |
| magnesium | 3327,0 | 0,6 | manganese | 11,1 | -0,4 |
| manganese | 11,1 | -0,4 | molybdenum | 0,0 | -0,4 |
| mercury | 0,0 | -0,4 | copper | 1,1 | -0,4 |
| molybdenum | 0,0 | -0,4 | selenium | 0,0 | -0,4 |
| nickel | 3,2 | 0,2 | arsenicum | 0,0 | -0,4 |
| phosphorus | 784,5 | -0,5 | lead | 0,0 | -0,4 |
| potassium | 14000,0 | -0,5 | mercury | 0,0 | -0,4 |
| selenium | 0,0 | -0,4 | lithium | 0,0 | -0,5 |
| sodium | 21350,0 | 5,7 | potassium | 14000,0 | -0,5 |
| vanadium | 1,3 | 1,2 | phosphorus | 784,5 | -0,5 |
| zinc | 46,4 | 1,7 | chromium | 0,0 | -0,6 |

# Carum carvi

| | |
|---|---|
| Family: | Umbelliferae |
| Scientific name: | Carum carvi |
| English name: | Caraway |
| German name: | KümmelCarvi |
| French name: | Cumin des pres |
| Dutch name: | Karwij |
| Part Used: | fructus |
| Extraction: | watery |
| Titer: | 1/5 |
| Standard minimal: | 1,45-2% etherische oil,8% carvon |
| Comment: | |

| Sorted by element name | | | Sorted by relative difference | | |
|---|---|---|---|---|---|
| aluminium | 8,3 | -0,3 | germanium | 1,0 | 0,8 |
| arsenicum | 0,0 | -0,4 | cobalt | 0,6 | 0,7 |
| cadmium | 0,1 | 0,0 | nickel | 3,0 | 0,1 |
| calcium | 816,2 | -0,5 | cadmium | 0,1 | 0,0 |
| chromium | 0,2 | -0,2 | phosphorus | 1502,0 | 0,0 |
| cobalt | 0,6 | 0,7 | chromium | 0,2 | -0,2 |
| copper | 3,0 | -0,2 | vanadium | 0,3 | -0,2 |
| germanium | 1,0 | 0,8 | iron | 35,5 | -0,2 |
| iron | 35,5 | -0,2 | copper | 3,0 | -0,2 |
| lead | 0,0 | -0,4 | aluminium | 8,3 | -0,3 |
| lithium | 0,0 | -0,5 | molybdenum | 0,0 | -0,4 |
| magnesium | 707,1 | -0,8 | potassium | 16450,0 | -0,4 |
| manganese | 5,7 | -0,4 | selenium | 0,0 | -0,4 |
| mercury | 0,0 | -0,4 | manganese | 5,7 | -0,4 |
| molybdenum | 0,0 | -0,4 | sodium | 665,7 | -0,4 |
| nickel | 3,0 | 0,1 | arsenicum | 0,0 | -0,4 |
| phosphorus | 1502,0 | 0,0 | lead | 0,0 | -0,4 |
| potassium | 16450,0 | -0,4 | mercury | 0,0 | -0,4 |
| selenium | 0,0 | -0,4 | lithium | 0,0 | -0,5 |
| sodium | 665,7 | -0,4 | calcium | 816,2 | -0,5 |
| vanadium | 0,3 | -0,2 | zinc | 4,6 | -0,7 |
| zinc | 4,6 | -0,7 | magnesium | 707,1 | -0,8 |

# Chamomilla

| | |
|---|---|
| Family: | Compositae |
| Scientific name: | Marticaria chamomilla |
| English name: | Chamomile |
| German name: | Echte Kamille |
| French name: | Chamomille commune |
| Dutch name: | Kamille |
| Part Used: | flores |
| Extraction: | alcoholic watery |
| Titer: | 1/5 |
| Standard minimal: | 0,3% bisabolol, 0,08% chamazulen |
| Comment: | |

| Sorted by element name | | | Sorted by relative difference | | |
|---|---|---|---|---|---|
| aluminium | 12,0 | -0,3 | selenium | 9,5 | 2,5 |
| arsenicum | 0,0 | -0,4 | sodium | 4400,0 | 0,7 |
| cadmium | 0,0 | -0,3 | lithium | 0,9 | 0,1 |
| calcium | 989,6 | -0,5 | copper | 6,0 | 0,0 |
| chromium | 0,2 | -0,2 | mercury | 0,2 | 0,0 |
| cobalt | 0,0 | -0,9 | chromium | 0,2 | -0,2 |
| copper | 6,0 | 0,0 | cadmium | 0,0 | -0,3 |
| germanium | 0,1 | -0,4 | aluminium | 12,0 | -0,3 |
| iron | 20,7 | -0,3 | potassium | 18500,0 | -0,3 |
| lead | 0,0 | -0,4 | nickel | 1,3 | -0,3 |
| lithium | 0,9 | 0,1 | iron | 20,7 | -0,3 |
| magnesium | 1061,0 | -0,6 | germanium | 0,1 | -0,4 |
| manganese | 9,9 | -0,4 | molybdenum | 0,0 | -0,4 |
| mercury | 0,2 | 0,0 | manganese | 9,9 | -0,4 |
| molybdenum | 0,0 | -0,4 | arsenicum | 0,0 | -0,4 |
| nickel | 1,3 | -0,3 | lead | 0,0 | -0,4 |
| phosphorus | 902,0 | -0,4 | phosphorus | 902,0 | -0,4 |
| potassium | 18500,0 | -0,3 | calcium | 989,6 | -0,5 |
| selenium | 9,5 | 2,5 | zinc | 8,4 | -0,5 |
| sodium | 4400,0 | 0,7 | magnesium | 1061,0 | -0,6 |
| vanadium | 0,0 | -0,7 | vanadium | 0,0 | -0,7 |
| zinc | 8,4 | -0,5 | cobalt | 0,0 | -0,9 |

# China officinalis

Family:              Rubiaceae
Scientific name:     Chinchona pubescens
English name:        Red chinchona
German name:         Chinabaum
French name:         Quinquina jaune
Dutch name:          Kinaboom
Part Used:           cortex
Extraction:          alcoholic watery
Titer:               1/4
Standard minimal:    10% kinin or total alkaloids
Comment:

| Sorted by element name | | | Sorted by relative difference | | |
|---|---|---|---|---|---|
| aluminium | 8,4 | -0,3 | chromium | 3,9 | 7,7 |
| arsenicum | 0,6 | -0,1 | lead | 2,7 | 2,3 |
| cadmium | 0,0 | -0,3 | mercury | 0,8 | 1,2 |
| calcium | 189,7 | -0,6 | cobalt | 0,3 | -0,1 |
| chromium | 3,9 | 7,7 | sodium | 1638,0 | -0,1 |
| cobalt | 0,3 | -0,1 | arsenicum | 0,6 | -0,1 |
| copper | 3,8 | -0,2 | iron | 46,5 | -0,2 |
| germanium | 0,0 | -0,5 | copper | 3,8 | -0,2 |
| iron | 46,5 | -0,2 | cadmium | 0,0 | -0,3 |
| lead | 2,7 | 2,3 | aluminium | 8,4 | -0,3 |
| lithium | 0,1 | -0,4 | molybdenum | 0,0 | -0,4 |
| magnesium | 50,6 | -1,1 | selenium | 0,0 | -0,4 |
| manganese | 1,3 | -0,4 | lithium | 0,1 | -0,4 |
| mercury | 0,8 | 1,2 | manganese | 1,3 | -0,4 |
| molybdenum | 0,0 | -0,4 | germanium | 0,0 | -0,5 |
| nickel | 0,0 | -0,6 | zinc | 7,2 | -0,5 |
| phosphorus | 206,1 | -0,9 | calcium | 189,7 | -0,6 |
| potassium | 1296,0 | -1,0 | nickel | 0,0 | -0,6 |
| selenium | 0,0 | -0,4 | vanadium | 0,0 | -0,7 |
| sodium | 1638,0 | -0,1 | phosphorus | 206,1 | -0,9 |
| vanadium | 0,0 | -0,7 | potassium | 1296,0 | -1,0 |
| zinc | 7,2 | -0,5 | magnesium | 50,6 | -1,1 |

# Chrysanthellum americanum

Family:
Scientific name:      Chrysanthellum americanum
English name:
German name:
French name:
Dutch name:
Part Used:      herba
Extraction:      watery
Titer:      1/5
Standard minimal:      1,5% chrysanthellin A+B
Comment:

| Sorted by element name | | | Sorted by relative difference | | |
|---|---|---|---|---|---|
| aluminium | 27,8 | 0,0 | germanium | 0,7 | 0,4 |
| arsenicum | 0,0 | -0,4 | phosphorus | 2143,0 | 0,4 |
| cadmium | 0,0 | -0,3 | potassium | 33060,0 | 0,3 |
| calcium | 2777,0 | -0,1 | iron | 110,0 | 0,3 |
| chromium | 0,3 | 0,1 | magnesium | 2708,0 | 0,3 |
| cobalt | 0,3 | -0,1 | chromium | 0,3 | 0,1 |
| copper | 2,5 | -0,3 | aluminium | 27,8 | 0,0 |
| germanium | 0,7 | 0,4 | cobalt | 0,3 | -0,1 |
| iron | 110,0 | 0,3 | zinc | 14,9 | -0,1 |
| lead | 0,0 | -0,4 | calcium | 2777,0 | -0,1 |
| lithium | 0,4 | -0,2 | vanadium | 0,4 | -0,1 |
| magnesium | 2708,0 | 0,3 | manganese | 41,7 | -0,2 |
| manganese | 41,7 | -0,2 | lithium | 0,4 | -0,2 |
| mercury | 0,0 | -0,4 | cadmium | 0,0 | -0,3 |
| molybdenum | 0,0 | -0,4 | copper | 2,5 | -0,3 |
| nickel | 0,0 | -0,6 | molybdenum | 0,0 | -0,4 |
| phosphorus | 2143,0 | 0,4 | selenium | 0,0 | -0,4 |
| potassium | 33060,0 | 0,3 | arsenicum | 0,0 | -0,4 |
| selenium | 0,0 | -0,4 | lead | 0,0 | -0,4 |
| sodium | 267,4 | -0,5 | mercury | 0,0 | -0,4 |
| vanadium | 0,4 | -0,1 | sodium | 267,4 | -0,5 |
| zinc | 14,9 | -0,1 | nickel | 0,0 | -0,6 |

# Chrysanthemum parthenium

Family:                 Compositae
Scientific name:        Chrysanthemum parthenium
English name:           Feverfew
German name:
French name:
Dutch name:             Moederkruid
Part Used:              herba
Extraction:             alcoholic watery
Titer:                  1/4
Standard minimal:
Comment:

| Sorted by element name | | | Sorted by relative difference | | |
|---|---|---|---|---|---|
| aluminium | 8,9 | -0,3 | cobalt | 1,0 | 1,7 |
| arsenicum | 0,0 | -0,4 | germanium | 0,7 | 0,4 |
| cadmium | 0,0 | -0,3 | copper | 5,4 | -0,1 |
| calcium | 352,3 | -0,6 | cadmium | 0,0 | -0,3 |
| chromium | 0,1 | -0,4 | aluminium | 8,9 | -0,3 |
| cobalt | 1,0 | 1,7 | mercury | 0,1 | -0,3 |
| copper | 5,4 | -0,1 | chromium | 0,1 | -0,4 |
| germanium | 0,7 | 0,4 | molybdenum | 0,0 | -0,4 |
| iron | 2,3 | -0,5 | selenium | 0,0 | -0,4 |
| lead | 0,0 | -0,4 | manganese | 3,6 | -0,4 |
| lithium | 0,0 | -0,5 | arsenicum | 0,0 | -0,4 |
| magnesium | 480,1 | -0,9 | lead | 0,0 | -0,4 |
| manganese | 3,6 | -0,4 | phosphorus | 894,7 | -0,5 |
| mercury | 0,1 | -0,3 | iron | 2,3 | -0,5 |
| molybdenum | 0,0 | -0,4 | lithium | 0,0 | -0,5 |
| nickel | 0,1 | -0,6 | sodium | 467,0 | -0,5 |
| phosphorus | 894,7 | -0,5 | zinc | 8,0 | -0,5 |
| potassium | 2802,0 | -0,9 | calcium | 352,3 | -0,6 |
| selenium | 0,0 | -0,4 | nickel | 0,1 | -0,6 |
| sodium | 467,0 | -0,5 | vanadium | 0,0 | -0,7 |
| vanadium | 0,0 | -0,7 | magnesium | 480,1 | -0,9 |
| zinc | 8,0 | -0,5 | potassium | 2802,0 | -0,9 |

# Crataegus oxyacantha

| | |
|---|---|
| Family: | Rosaceae |
| Scientific name: | Crataegus oxyacantha |
| English name: | Hawthorn |
| German name: | Weissdorn |
| French name: | Aubépine |
| Dutch name: | Meidoorn |
| Part Used: | folia cum flores |
| Extraction: | alcoholic watery |
| Titer: | 1/5 |
| Standard minimal: | 1% vitexin -O- rhmanosid |
| Comment: | |

| Sorted by element name | | | Sorted by relative difference | | |
|---|---|---|---|---|---|
| aluminium | 0,0 | -0,5 | nickel | 7,0 | 1,1 |
| arsenicum | 0,0 | -0,4 | lithium | 1,7 | 0,6 |
| cadmium | 0,0 | -0,3 | copper | 13,2 | 0,5 |
| calcium | 1397,0 | -0,4 | zinc | 24,2 | 0,4 |
| chromium | 0,1 | -0,4 | mercury | 0,3 | 0,2 |
| cobalt | 0,0 | -0,9 | sodium | 2008,0 | 0,0 |
| copper | 13,2 | 0,5 | potassium | 24810,0 | 0,0 |
| germanium | 0,0 | -0,5 | magnesium | 2081,0 | -0,1 |
| iron | 8,3 | -0,4 | phosphorus | 1242,0 | -0,2 |
| lead | 0,0 | -0,4 | cadmium | 0,0 | -0,3 |
| lithium | 1,7 | 0,6 | chromium | 0,1 | -0,4 |
| magnesium | 2081,0 | -0,1 | molybdenum | 0,0 | -0,4 |
| manganese | 6,8 | -0,4 | calcium | 1397,0 | -0,4 |
| mercury | 0,3 | 0,2 | manganese | 6,8 | -0,4 |
| molybdenum | 0,0 | -0,4 | selenium | 0,0 | -0,4 |
| nickel | 7,0 | 1,1 | iron | 8,3 | -0,4 |
| phosphorus | 1242,0 | -0,2 | arsenicum | 0,0 | -0,4 |
| potassium | 24810,0 | 0,0 | lead | 0,0 | -0,4 |
| selenium | 0,0 | -0,4 | aluminium | 0,0 | -0,5 |
| sodium | 2008,0 | 0,0 | germanium | 0,0 | -0,5 |
| vanadium | 0,0 | -0,7 | vanadium | 0,0 | -0,7 |
| zinc | 24,2 | 0,4 | cobalt | 0,0 | -0,9 |

# Cupressus sempervirens

| | |
|---|---|
| Family: | Cupressaceae |
| Scientific name: | Cupressus sempervirens |
| English name: | Cyper |
| German name: | Zypresse |
| French name: | Cipres |
| Dutch name: | Cypres |
| Part Used: | fructus |
| Extraction: | watery |
| Titer: | 1/8 |
| Standard minimal: | |
| Comment: | |

| Sorted by element name | | | Sorted by relative difference | | |
|---|---|---|---|---|---|
| aluminium | 105,4 | 1,1 | phosphorus | 3545,0 | 1,4 |
| arsenicum | 0,0 | -0,4 | potassium | 52760,0 | 1,2 |
| cadmium | 0,0 | -0,3 | aluminium | 105,4 | 1,1 |
| calcium | 4062,0 | 0,2 | chromium | 0,7 | 0,9 |
| chromium | 0,7 | 0,9 | cobalt | 0,6 | 0,7 |
| cobalt | 0,6 | 0,7 | lithium | 1,7 | 0,6 |
| copper | 3,3 | -0,2 | iron | 146,5 | 0,5 |
| germanium | 0,0 | -0,5 | sodium | 2890,0 | 0,3 |
| iron | 146,5 | 0,5 | vanadium | 0,6 | 0,2 |
| lead | 0,0 | -0,4 | calcium | 4062,0 | 0,2 |
| lithium | 1,7 | 0,6 | magnesium | 1987,0 | -0,1 |
| magnesium | 1987,0 | -0,1 | zinc | 13,7 | -0,2 |
| manganese | 30,5 | -0,2 | copper | 3,3 | -0,2 |
| mercury | 0,0 | -0,4 | manganese | 30,5 | -0,2 |
| molybdenum | 0,0 | -0,4 | cadmium | 0,0 | -0,3 |
| nickel | 0,0 | -0,6 | molybdenum | 0,0 | -0,4 |
| phosphorus | 3545,0 | 1,4 | selenium | 0,0 | -0,4 |
| potassium | 52760,0 | 1,2 | arsenicum | 0,0 | -0,4 |
| selenium | 0,0 | -0,4 | lead | 0,0 | -0,4 |
| sodium | 2890,0 | 0,3 | mercury | 0,0 | -0,4 |
| vanadium | 0,6 | 0,2 | germanium | 0,0 | -0,5 |
| zinc | 13,7 | -0,2 | nickel | 0,0 | -0,6 |

# Curcuma

| | |
|---|---|
| Family: | Zingiberaceae |
| Scientific name: | Curcuma |
| English name: | Temoe lawak |
| German name: | |
| French name: | Curcuma |
| Dutch name: | Kurkuma |
| Part Used: | rhizoma |
| Extraction: | alcoholic watery |
| Titer: | 1/4 |
| Standard minimal: | 1% curcumin |
| Comment: | |

| Sorted by element name | | | Sorted by relative difference | | |
|---|---|---|---|---|---|
| aluminium | 96,0 | 1,0 | chromium | 1,3 | 2,2 |
| arsenicum | 0,0 | -0,4 | potassium | 49690,0 | 1,0 |
| cadmium | 0,0 | -0,3 | aluminium | 96,0 | 1,0 |
| calcium | 974,1 | -0,5 | phosphorus | 2037,0 | 0,3 |
| chromium | 1,3 | 2,2 | copper | 8,8 | 0,2 |
| cobalt | 0,0 | -0,9 | zinc | 16,3 | 0,0 |
| copper | 8,8 | 0,2 | manganese | 60,3 | 0,0 |
| germanium | 0,0 | -0,5 | iron | 58,9 | -0,1 |
| iron | 58,9 | -0,1 | magnesium | 1868,0 | -0,2 |
| lead | 0,0 | -0,4 | cadmium | 0,0 | -0,3 |
| lithium | 0,0 | -0,5 | molybdenum | 0,0 | -0,4 |
| magnesium | 1868,0 | -0,2 | selenium | 0,0 | -0,4 |
| manganese | 60,3 | 0,0 | arsenicum | 0,0 | -0,4 |
| mercury | 0,0 | -0,4 | lead | 0,0 | -0,4 |
| molybdenum | 0,0 | -0,4 | mercury | 0,0 | -0,4 |
| nickel | 0,0 | -0,6 | calcium | 974,1 | -0,5 |
| phosphorus | 2037,0 | 0,3 | lithium | 0,0 | -0,5 |
| potassium | 49690,0 | 1,0 | germanium | 0,0 | -0,5 |
| selenium | 0,0 | -0,4 | sodium | 368,8 | -0,5 |
| sodium | 368,8 | -0,5 | nickel | 0,0 | -0,6 |
| vanadium | 0,0 | -0,7 | vanadium | 0,0 | -0,7 |
| zinc | 16,3 | 0,0 | cobalt | 0,0 | -0,9 |

# Cynara scolymus

| | |
|---|---|
| Family: | Compositae |
| Scientific name: | Cynara scolymus |
| English name: | Artichoke |
| German name: | Echte artichocke |
| French name: | Artichaut |
| Dutch name: | Artisjok |
| Part Used: | folia |
| Extraction: | watery |
| Titer: | 1/4 |
| Standard minimal: | 1% cynarin |
| Comment: | |

| Sorted by element name | | | Sorted by relative difference | | |
|---|---|---|---|---|---|
| aluminium | 7,7 | -0,3 | sodium | 3338,0 | 0,4 |
| arsenicum | 0,9 | 0,0 | cadmium | 0,1 | 0,1 |
| cadmium | 0,1 | 0,1 | arsenicum | 0,9 | 0,0 |
| calcium | 1382,0 | -0,4 | chromium | 0,3 | 0,0 |
| chromium | 0,3 | 0,0 | iron | 23,7 | -0,3 |
| cobalt | 0,0 | -0,9 | aluminium | 7,7 | -0,3 |
| copper | 0,0 | -0,5 | molybdenum | 0,0 | -0,4 |
| germanium | 0,0 | -0,5 | calcium | 1382,0 | -0,4 |
| iron | 23,7 | -0,3 | selenium | 0,0 | -0,4 |
| lead | 0,0 | -0,4 | manganese | 4,4 | -0,4 |
| lithium | 0,0 | -0,5 | lead | 0,0 | -0,4 |
| magnesium | 456,6 | -0,9 | mercury | 0,0 | -0,4 |
| manganese | 4,4 | -0,4 | copper | 0,0 | -0,5 |
| mercury | 0,0 | -0,4 | lithium | 0,0 | -0,5 |
| molybdenum | 0,0 | -0,4 | potassium | 14310,0 | -0,5 |
| nickel | 0,3 | -0,6 | germanium | 0,0 | -0,5 |
| phosphorus | 540,0 | -0,7 | nickel | 0,3 | -0,6 |
| potassium | 14310,0 | -0,5 | phosphorus | 540,0 | -0,7 |
| selenium | 0,0 | -0,4 | zinc | 4,2 | -0,7 |
| sodium | 3338,0 | 0,4 | vanadium | 0,0 | -0,7 |
| vanadium | 0,0 | -0,7 | cobalt | 0,0 | -0,9 |
| zinc | 4,2 | -0,7 | magnesium | 456,6 | -0,9 |

# Echinacea angustifolia

Family:              Compositae
Scientific name:     Echinacea angustifolia
English name:        Nigerhead
German name:         Schmalblättrige Kegelblume
French name:
Dutch name:          Smalbladige zonnehoed
Part Used:           radix
Extraction:          alcoholic watery
Titer:               1/6,5
Standard minimal:    0,3% echinacosid
Comment:

| Sorted by element name | | | Sorted by relative difference | | |
|---|---|---|---|---|---|
| aluminium | 5,5 | -0,4 | phosphorus | 3763,0 | 1,5 |
| arsenicum | 0,0 | -0,4 | copper | 18,6 | 0,9 |
| cadmium | 0,0 | -0,3 | chromium | 0,5 | 0,5 |
| calcium | 717,9 | -0,5 | sodium | 1995,0 | 0,0 |
| chromium | 0,5 | 0,5 | zinc | 13,7 | -0,2 |
| cobalt | 0,0 | -0,9 | cadmium | 0,0 | -0,3 |
| copper | 18,6 | 0,9 | magnesium | 1597,0 | -0,3 |
| germanium | 0,0 | -0,5 | nickel | 1,3 | -0,3 |
| iron | 8,5 | -0,4 | molybdenum | 0,0 | -0,4 |
| lead | 0,0 | -0,4 | aluminium | 5,5 | -0,4 |
| lithium | 0,0 | -0,5 | selenium | 0,0 | -0,4 |
| magnesium | 1597,0 | -0,3 | manganese | 3,3 | -0,4 |
| manganese | 3,3 | -0,4 | iron | 8,5 | -0,4 |
| mercury | 0,0 | -0,4 | arsenicum | 0,0 | -0,4 |
| molybdenum | 0,0 | -0,4 | lead | 0,0 | -0,4 |
| nickel | 1,3 | -0,3 | mercury | 0,0 | -0,4 |
| phosphorus | 3763,0 | 1,5 | lithium | 0,0 | -0,5 |
| potassium | 14200,0 | -0,5 | potassium | 14200,0 | -0,5 |
| selenium | 0,0 | -0,4 | germanium | 0,0 | -0,5 |
| sodium | 1995,0 | 0,0 | calcium | 717,9 | -0,5 |
| vanadium | 0,0 | -0,7 | vanadium | 0,0 | -0,7 |
| zinc | 13,7 | -0,2 | cobalt | 0,0 | -0,9 |

# Echinacea purpurea

| | |
|---|---|
| Family: | Compositae |
| Scientific name: | Echinacea purpurea |
| English name: | Black Samson |
| German name: | Purpurfarbene Kegelblume |
| French name: | |
| Dutch name: | Purperen zonnehoed |
| Part Used: | radix |
| Extraction: | alcoholic watery |
| Titer: | 1/6 |
| Standard minimal: | |
| Comment: | |

| Sorted by element name | | | Sorted by relative difference | | |
|---|---|---|---|---|---|
| aluminium | 5,3 | -0,4 | selenium | 10,5 | 2,8 |
| arsenicum | 0,0 | -0,4 | mercury | 0,8 | 1,2 |
| cadmium | 0,1 | 0,1 | magnesium | 3040,0 | 0,4 |
| calcium | 1980,0 | -0,3 | potassium | 30340,0 | 0,2 |
| chromium | 0,3 | 0,1 | sodium | 2412,0 | 0,1 |
| cobalt | 0,0 | -0,9 | cadmium | 0,1 | 0,1 |
| copper | 2,1 | -0,3 | chromium | 0,3 | 0,1 |
| germanium | 0,0 | -0,5 | phosphorus | 1291,0 | -0,2 |
| iron | 11,7 | -0,4 | zinc | 13,1 | -0,2 |
| lead | 0,0 | -0,4 | calcium | 1980,0 | -0,3 |
| lithium | 0,2 | -0,3 | copper | 2,1 | -0,3 |
| magnesium | 3040,0 | 0,4 | lithium | 0,2 | -0,3 |
| manganese | 6,6 | -0,4 | molybdenum | 0,0 | -0,4 |
| mercury | 0,8 | 1,2 | aluminium | 5,3 | -0,4 |
| molybdenum | 0,0 | -0,4 | iron | 11,7 | -0,4 |
| nickel | 0,4 | -0,5 | manganese | 6,6 | -0,4 |
| phosphorus | 1291,0 | -0,2 | arsenicum | 0,0 | -0,4 |
| potassium | 30340,0 | 0,2 | lead | 0,0 | -0,4 |
| selenium | 10,5 | 2,8 | germanium | 0,0 | -0,5 |
| sodium | 2412,0 | 0,1 | nickel | 0,4 | -0,5 |
| vanadium | 0,0 | -0,7 | vanadium | 0,0 | -0,7 |
| zinc | 13,1 | -0,2 | cobalt | 0,0 | -0,9 |

# Eleutherococcus

| | |
|---|---|
| Family: | Araliaceae |
| Scientific name: | Eleutherococcus senticosus |
| English name: | Siberic ginseng |
| German name: | Taira wurzel |
| French name: | Eleutherocoque |
| Dutch name: | Russische ginseng |
| Part Used: | radix |
| Extraction: | alcoholic watery |
| Titer: | 1/5 |
| Standard minimal: | 0,2% eleutherosid B, 0,2% |
| eleutherosid E | Comment: |

| Sorted by element name | | | Sorted by relative difference | | |
|---|---|---|---|---|---|
| aluminium | 6,4 | -0,4 | mercury | 0,3 | 0,2 |
| arsenicum | 0,0 | -0,4 | germanium | 0,3 | -0,1 |
| cadmium | 0,0 | -0,3 | cadmium | 0,0 | -0,3 |
| calcium | 654,9 | -0,5 | nickel | 1,3 | -0,3 |
| chromium | 0,0 | -0,6 | copper | 1,9 | -0,3 |
| cobalt | 0,0 | -0,9 | aluminium | 6,4 | -0,4 |
| copper | 1,9 | -0,3 | molybdenum | 0,0 | -0,4 |
| germanium | 0,3 | -0,1 | manganese | 7,0 | -0,4 |
| iron | 4,5 | -0,4 | selenium | 0,0 | -0,4 |
| lead | 0,0 | -0,4 | lithium | 0,1 | -0,4 |
| lithium | 0,1 | -0,4 | phosphorus | 942,4 | -0,4 |
| magnesium | 781,2 | -0,7 | arsenicum | 0,0 | -0,4 |
| manganese | 7,0 | -0,4 | iron | 4,5 | -0,4 |
| mercury | 0,3 | 0,2 | lead | 0,0 | -0,4 |
| molybdenum | 0,0 | -0,4 | calcium | 654,9 | -0,5 |
| nickel | 1,3 | -0,3 | sodium | 172,2 | -0,5 |
| phosphorus | 942,4 | -0,4 | chromium | 0,0 | -0,6 |
| potassium | 8564,0 | -0,7 | potassium | 8564,0 | -0,7 |
| selenium | 0,0 | -0,4 | zinc | 4,3 | -0,7 |
| sodium | 172,2 | -0,5 | magnesium | 781,2 | -0,7 |
| vanadium | 0,0 | -0,7 | vanadium | 0,0 | -0,7 |
| zinc | 4,3 | -0,7 | cobalt | 0,0 | -0,9 |

# Equisetum arvense

| | |
|---|---|
| Family: | Equisetaceae |
| Scientific name: | Equisetum arvense |
| English name: | Shave grass, Horsetail |
| German name: | Ackerschachtelhalm |
| French name: | Prêle des champs |
| Dutch name: | Paardestaart |
| Part Used: | herba |
| Extraction: | watery |
| Titer: | 1/5 |
| Standard minimal: | |
| Comment: | |

| Sorted by element name | | | Sorted by relative difference | | |
|---|---|---|---|---|---|
| aluminium | 4,7 | -0,4 | magnesium | 3539,0 | 0,7 |
| arsenicum | 0,0 | -0,4 | cobalt | 0,5 | 0,4 |
| cadmium | 0,0 | -0,3 | potassium | 28470,0 | 0,1 |
| calcium | 1747,0 | -0,3 | chromium | 0,3 | 0,1 |
| chromium | 0,3 | 0,1 | sodium | 1236,0 | -0,2 |
| cobalt | 0,5 | 0,4 | cadmium | 0,0 | -0,3 |
| copper | 0,8 | -0,4 | calcium | 1747,0 | -0,3 |
| germanium | 0,0 | -0,5 | manganese | 13,6 | -0,3 |
| iron | 13,7 | -0,4 | molybdenum | 0,0 | -0,4 |
| lead | 0,0 | -0,4 | iron | 13,7 | -0,4 |
| lithium | 0,0 | -0,5 | aluminium | 4,7 | -0,4 |
| magnesium | 3539,0 | 0,7 | selenium | 0,0 | -0,4 |
| manganese | 13,6 | -0,3 | copper | 0,8 | -0,4 |
| mercury | 0,0 | -0,4 | nickel | 0,8 | -0,4 |
| molybdenum | 0,0 | -0,4 | arsenicum | 0,0 | -0,4 |
| nickel | 0,8 | -0,4 | lead | 0,0 | -0,4 |
| phosphorus | 841,5 | -0,5 | mercury | 0,0 | -0,4 |
| potassium | 28470,0 | 0,1 | lithium | 0,0 | -0,5 |
| selenium | 0,0 | -0,4 | germanium | 0,0 | -0,5 |
| sodium | 1236,0 | -0,2 | phosphorus | 841,5 | -0,5 |
| vanadium | 0,0 | -0,7 | zinc | 4,4 | -0,7 |
| zinc | 4,4 | -0,7 | vanadium | 0,0 | -0,7 |

# Erigeron canadensis

| | |
|---|---|
| Family: | Compositae |
| Scientific name: | Erigeron canadensis |
| English name: | Horseweed, Canada fleabean |
| German name: | Pferdeunkraut |
| French name: | Vergerette du Canada |
| Dutch name: | Fijnstraal canadees |
| Part Used: | herba |
| Extraction: | alcoholic watery |
| Titer: | 1/10 |
| Standard minimal: | |
| Comment: | |

| Sorted by element name | | | Sorted by relative difference | | |
|---|---|---|---|---|---|
| aluminium | 0,0 | -0,5 | copper | 7,5 | 0,1 |
| arsenicum | 0,0 | -0,4 | mercury | 0,2 | 0,0 |
| cadmium | 0,0 | -0,3 | cadmium | 0,0 | -0,3 |
| calcium | 515,4 | -0,6 | lithium | 0,3 | -0,3 |
| chromium | 0,0 | -0,6 | potassium | 18840,0 | -0,3 |
| cobalt | 0,0 | -0,9 | molybdenum | 0,0 | -0,4 |
| copper | 7,5 | 0,1 | phosphorus | 1028,0 | -0,4 |
| germanium | 0,0 | -0,5 | manganese | 5,9 | -0,4 |
| iron | 4,7 | -0,4 | selenium | 0,0 | -0,4 |
| lead | 0,0 | -0,4 | zinc | 9,1 | -0,4 |
| lithium | 0,3 | -0,3 | arsenicum | 0,0 | -0,4 |
| magnesium | 413,6 | -0,9 | iron | 4,7 | -0,4 |
| manganese | 5,9 | -0,4 | lead | 0,0 | -0,4 |
| mercury | 0,2 | 0,0 | aluminium | 0,0 | -0,5 |
| molybdenum | 0,0 | -0,4 | germanium | 0,0 | -0,5 |
| nickel | 0,1 | -0,6 | calcium | 515,4 | -0,6 |
| phosphorus | 1028,0 | -0,4 | sodium | 109,0 | -0,6 |
| potassium | 18840,0 | -0,3 | chromium | 0,0 | -0,6 |
| selenium | 0,0 | -0,4 | nickel | 0,1 | -0,6 |
| sodium | 109,0 | -0,6 | vanadium | 0,0 | -0,7 |
| vanadium | 0,0 | -0,7 | cobalt | 0,0 | -0,9 |
| zinc | 9,1 | -0,4 | magnesium | 413,6 | -0,9 |

# Eschscholtzia californica

Family:                 Papaveraceae
Scientific name:        Eschscholtzia californica
English name:           Californian poppy
German name:            Kalifornischer Kappen-Mohn
French name:            Globe du soleil
Dutch name:             Goudpapaver, slaapmutsje
Part Used:              herba
Extraction:             watery
Titer:                  1/5
Standard minimal:
Comment:

| Sorted by element name | | | Sorted by relative difference | | |
|---|---|---|---|---|---|
| aluminium | 125,6 | 1,5 | zinc | 84,3 | 3,9 |
| arsenicum | 2,6 | 0,9 | potassium | 85830,0 | 2,5 |
| cadmium | 0,0 | -0,3 | phosphorus | 4436,0 | 2,0 |
| calcium | 2478,0 | -0,2 | iron | 285,2 | 1,5 |
| chromium | 0,7 | 0,9 | aluminium | 125,6 | 1,5 |
| cobalt | 0,5 | 0,4 | magnesium | 4591,0 | 1,2 |
| copper | 8,4 | 0,2 | chromium | 0,7 | 0,9 |
| germanium | 0,0 | -0,5 | arsenicum | 2,6 | 0,9 |
| iron | 285,2 | 1,5 | lithium | 1,8 | 0,6 |
| lead | 0,0 | -0,4 | cobalt | 0,5 | 0,4 |
| lithium | 1,8 | 0,6 | manganese | 120,7 | 0,4 |
| magnesium | 4591,0 | 1,2 | mercury | 0,4 | 0,4 |
| manganese | 120,7 | 0,4 | copper | 8,4 | 0,2 |
| mercury | 0,4 | 0,4 | vanadium | 0,5 | 0,0 |
| molybdenum | 0,0 | -0,4 | calcium | 2478,0 | -0,2 |
| nickel | 0,4 | -0,5 | cadmium | 0,0 | -0,3 |
| phosphorus | 4436,0 | 2,0 | sodium | 1118,0 | -0,3 |
| potassium | 85830,0 | 2,5 | molybdenum | 0,0 | -0,4 |
| selenium | 0,0 | -0,4 | selenium | 0,0 | -0,4 |
| sodium | 1118,0 | -0,3 | lead | 0,0 | -0,4 |
| vanadium | 0,5 | 0,0 | germanium | 0,0 | -0,5 |
| zinc | 84,3 | 3,9 | nickel | 0,4 | -0,5 |

# Eucalyptus

| | |
|---|---|
| Family: | Myrtaceae |
| Scientific name: | Eucalyptus |
| English name: | Blue gum-tree |
| German name: | Blaugummibaum, Fieberbaum |
| French name: | Arbre de la fièvre |
| Dutch name: | Eucalyptus |
| Part Used: | folia |
| Extraction: | watery |
| Titer: | 1/5 |
| Standard minimal: | 3% etheric oil |
| Comment: | |

| Sorted by element name | | | Sorted by relative difference | | |
|---|---|---|---|---|---|
| aluminium | 8,6 | -0,3 | manganese | 909,6 | 5,7 |
| arsenicum | 0,0 | -0,4 | lithium | 1,6 | 0,5 |
| cadmium | 0,1 | 0,1 | lead | 0,9 | 0,5 |
| calcium | 1005,0 | -0,5 | sodium | 2433,0 | 0,1 |
| chromium | 0,1 | -0,4 | nickel | 3,0 | 0,1 |
| cobalt | 0,3 | -0,1 | cadmium | 0,1 | 0,1 |
| copper | 0,9 | -0,4 | cobalt | 0,3 | -0,1 |
| germanium | 0,0 | -0,5 | vanadium | 0,3 | -0,3 |
| iron | 1,8 | -0,5 | aluminium | 8,6 | -0,3 |
| lead | 0,9 | 0,5 | chromium | 0,1 | -0,4 |
| lithium | 1,6 | 0,5 | molybdenum | 0,0 | -0,4 |
| magnesium | 1236,0 | -0,5 | copper | 0,9 | -0,4 |
| manganese | 909,6 | 5,7 | selenium | 0,0 | -0,4 |
| mercury | 0,0 | -0,4 | arsenicum | 0,0 | -0,4 |
| molybdenum | 0,0 | -0,4 | mercury | 0,0 | -0,4 |
| nickel | 3,0 | 0,1 | calcium | 1005,0 | -0,5 |
| phosphorus | 391,6 | -0,8 | iron | 1,8 | -0,5 |
| potassium | 8740,0 | -0,7 | germanium | 0,0 | -0,5 |
| selenium | 0,0 | -0,4 | magnesium | 1236,0 | -0,5 |
| sodium | 2433,0 | 0,1 | zinc | 6,1 | -0,6 |
| vanadium | 0,3 | -0,3 | potassium | 8740,0 | -0,7 |
| zinc | 6,1 | -0,6 | phosphorus | 391,6 | -0,8 |

# Eugenia caryophyllata

| | |
|---|---|
| Family: | Myrtaceae |
| Scientific name: | Eugenia caryophyllata |
| English name: | Cloves, Rose-apple |
| German name: | |
| French name: | Giroflier |
| Dutch name: | Kruidnagel |
| Part Used: | flores |
| Extraction: | alcoholic watery |
| Titer: | 1/3 |
| Standard minimal: | 1% eugenol |
| Comment: | |

| Sorted by element name | | | Sorted by relative difference | | |
|---|---|---|---|---|---|
| aluminium | 0,0 | -0,5 | sodium | 2253,0 | 0,1 |
| arsenicum | 0,0 | -0,4 | cadmium | 0,0 | -0,3 |
| cadmium | 0,0 | -0,3 | molybdenum | 0,0 | -0,4 |
| calcium | 344,7 | -0,6 | manganese | 5,8 | -0,4 |
| chromium | 0,0 | -0,6 | selenium | 0,0 | -0,4 |
| cobalt | 0,1 | -0,6 | iron | 11,0 | -0,4 |
| copper | 0,2 | -0,5 | arsenicum | 0,0 | -0,4 |
| germanium | 0,0 | -0,5 | lead | 0,0 | -0,4 |
| iron | 11,0 | -0,4 | mercury | 0,0 | -0,4 |
| lead | 0,0 | -0,4 | copper | 0,2 | -0,5 |
| lithium | 0,0 | -0,5 | aluminium | 0,0 | -0,5 |
| magnesium | 184,4 | -1,0 | lithium | 0,0 | -0,5 |
| manganese | 5,8 | -0,4 | germanium | 0,0 | -0,5 |
| mercury | 0,0 | -0,4 | chromium | 0,0 | -0,6 |
| molybdenum | 0,0 | -0,4 | calcium | 344,7 | -0,6 |
| nickel | 0,0 | -0,6 | cobalt | 0,1 | -0,6 |
| phosphorus | 236,6 | -0,9 | nickel | 0,0 | -0,6 |
| potassium | 3255,0 | -0,9 | vanadium | 0,0 | -0,7 |
| selenium | 0,0 | -0,4 | zinc | 2,4 | -0,8 |
| sodium | 2253,0 | 0,1 | phosphorus | 236,6 | -0,9 |
| vanadium | 0,0 | -0,7 | potassium | 3255,0 | -0,9 |
| zinc | 2,4 | -0,8 | magnesium | 184,4 | -1,0 |

# Fenugrec

| | |
|---|---|
| Family: | Umbelliferae |
| Scientific name: | Trigonella foenum-graecum |
| English name: | Fenugreek |
| German name: | Backshornklee |
| French name: | Fenugrec |
| Dutch name: | Fenegriek |
| Part Used: | semen |
| Extraction: | alcoholic watery |
| Titer: | 1/5 |
| Standard minimal: | 3% vitexin |
| Comment: | |

| Sorted by element name | | | Sorted by relative difference | | |
|---|---|---|---|---|---|
| aluminium | 6,9 | -0,4 | copper | 5,4 | -0,1 |
| arsenicum | 0,0 | -0,4 | nickel | 1,6 | -0,2 |
| cadmium | 0,0 | -0,3 | cadmium | 0,0 | -0,3 |
| calcium | 254,0 | -0,6 | iron | 18,9 | -0,3 |
| chromium | 0,0 | -0,6 | aluminium | 6,9 | -0,4 |
| cobalt | 0,0 | -0,9 | molybdenum | 0,0 | -0,4 |
| copper | 5,4 | -0,1 | selenium | 0,0 | -0,4 |
| germanium | 0,0 | -0,5 | lithium | 0,1 | -0,4 |
| iron | 18,9 | -0,3 | manganese | 1,5 | -0,4 |
| lead | 0,0 | -0,4 | arsenicum | 0,0 | -0,4 |
| lithium | 0,1 | -0,4 | lead | 0,0 | -0,4 |
| magnesium | 512,0 | -0,9 | mercury | 0,0 | -0,4 |
| manganese | 1,5 | -0,4 | germanium | 0,0 | -0,5 |
| mercury | 0,0 | -0,4 | sodium | 258,9 | -0,5 |
| molybdenum | 0,0 | -0,4 | chromium | 0,0 | -0,6 |
| nickel | 1,6 | -0,2 | calcium | 254,0 | -0,6 |
| phosphorus | 439,9 | -0,8 | zinc | 5,1 | -0,7 |
| potassium | 8571,0 | -0,7 | potassium | 8571,0 | -0,7 |
| selenium | 0,0 | -0,4 | vanadium | 0,0 | -0,7 |
| sodium | 258,9 | -0,5 | phosphorus | 439,9 | -0,8 |
| vanadium | 0,0 | -0,7 | cobalt | 0,0 | -0,9 |
| zinc | 5,1 | -0,7 | magnesium | 512,0 | -0,9 |

# Foeniculum vulgare

| | |
|---|---|
| Family: | Umbelliferae |
| Scientific name: | Foeniculum vulgare |
| English name: | Fennel |
| German name: | Fenchel |
| French name: | Fenonil |
| Dutch name: | Venkel |
| Part Used: | fructus |
| Extraction: | watery |
| Titer: | 1/4,5 |
| Standard minimal: | 3,5% etheric oil |
| Comment: | |

| Sorted by element name | | | Sorted by relative difference | | |
|---|---|---|---|---|---|
| aluminium | 21,3 | -0,1 | lithium | 5,2 | 2,7 |
| arsenicum | 0,0 | -0,4 | selenium | 8,8 | 2,3 |
| cadmium | 0,0 | -0,3 | molybdenum | 1,4 | 1,7 |
| calcium | 2134,0 | -0,2 | magnesium | 4327,0 | 1,1 |
| chromium | 0,2 | -0,2 | sodium | 5136,0 | 0,9 |
| cobalt | 0,0 | -0,9 | potassium | 45360,0 | 0,8 |
| copper | 13,2 | 0,5 | copper | 13,2 | 0,5 |
| germanium | 0,0 | -0,5 | phosphorus | 1681,0 | 0,1 |
| iron | 31,1 | -0,3 | zinc | 17,0 | 0,0 |
| lead | 0,3 | -0,1 | nickel | 2,4 | 0,0 |
| lithium | 5,2 | 2,7 | aluminium | 21,3 | -0,1 |
| magnesium | 4327,0 | 1,1 | lead | 0,3 | -0,1 |
| manganese | 7,3 | -0,4 | chromium | 0,2 | -0,2 |
| mercury | 0,0 | -0,4 | calcium | 2134,0 | -0,2 |
| molybdenum | 1,4 | 1,7 | cadmium | 0,0 | -0,3 |
| nickel | 2,4 | 0,0 | iron | 31,1 | -0,3 |
| phosphorus | 1681,0 | 0,1 | manganese | 7,3 | -0,4 |
| potassium | 45360,0 | 0,8 | arsenicum | 0,0 | -0,4 |
| selenium | 8,8 | 2,3 | mercury | 0,0 | -0,4 |
| sodium | 5136,0 | 0,9 | germanium | 0,0 | -0,5 |
| vanadium | 0,0 | -0,7 | vanadium | 0,0 | -0,7 |
| zinc | 17,0 | 0,0 | cobalt | 0,0 | -0,9 |

# Fraxinus exceliors

Family:                Oleaceae
Scientific name:       Fraxinus excelsior
English name:          Common ash
German name:           Gemeine Esche
French name:           Frêne élevé
Dutch name:            Es
Part Used:             folia
Extraction:            watery
Titer:                 1/4
Standard minimal:      0,3% esculosid
Comment:

| Sorted by element name | | | Sorted by relative difference | | |
|---|---|---|---|---|---|
| aluminium | 39,0 | 0,1 | calcium | 26480,0 | 4,7 |
| arsenicum | 5,1 | 2,2 | mercury | 2,5 | 4,6 |
| cadmium | 0,0 | -0,3 | lead | 3,2 | 2,8 |
| calcium | 26480,0 | 4,7 | magnesium | 6586,0 | 2,2 |
| chromium | 0,0 | -0,6 | arsenicum | 5,1 | 2,2 |
| cobalt | 0,7 | 1,0 | vanadium | 1,7 | 1,8 |
| copper | 1,4 | -0,4 | phosphorus | 3580,0 | 1,4 |
| germanium | 0,0 | -0,5 | cobalt | 0,7 | 1,0 |
| iron | 205,8 | 0,9 | iron | 205,8 | 0,9 |
| lead | 3,2 | 2,8 | lithium | 2,1 | 0,8 |
| lithium | 2,1 | 0,8 | potassium | 36630,0 | 0,5 |
| magnesium | 6586,0 | 2,2 | aluminium | 39,0 | 0,1 |
| manganese | 60,3 | 0,0 | sodium | 2287,0 | 0,1 |
| mercury | 2,5 | 4,6 | nickel | 2,8 | 0,1 |
| molybdenum | 0,0 | -0,4 | manganese | 60,3 | 0,0 |
| nickel | 2,8 | 0,1 | cadmium | 0,0 | -0,3 |
| phosphorus | 3580,0 | 1,4 | copper | 1,4 | -0,4 |
| potassium | 36630,0 | 0,5 | molybdenum | 0,0 | -0,4 |
| selenium | 0,0 | -0,4 | selenium | 0,0 | -0,4 |
| sodium | 2287,0 | 0,1 | zinc | 8,4 | -0,5 |
| vanadium | 1,7 | 1,8 | germanium | 0,0 | -0,5 |
| zinc | 8,4 | -0,5 | chromium | 0,0 | -0,6 |

# Fucus vesiculosus

| | |
|---|---|
| Family: | Fucaceae |
| Scientific name: | Fucus vesiculosus |
| English name: | Bladder wrack, Sea weed |
| German name: | Blasentang |
| French name: | Varech |
| Dutch name: | Blaaswier |
| Part Used: | thallus |
| Extraction: | watery |
| Titer: | 1/6 |
| Standard minimal: | 0,1-0,2& iodum |
| Comment: | it looks that fucus must have much |

in common with natrum arsenicisum

| Sorted by element name | | | Sorted by relative difference | | |
|---|---|---|---|---|---|
| aluminium | 8,1 | -0,3 | sodium | 21220,0 | 5,7 |
| arsenicum | 9,2 | 4,3 | arsenicum | 9,2 | 4,3 |
| cadmium | 0,0 | -0,3 | magnesium | 2657,0 | 0,2 |
| calcium | 1426,0 | -0,4 | chromium | 0,2 | -0,2 |
| chromium | 0,2 | -0,2 | lithium | 0,4 | -0,2 |
| cobalt | 0,0 | -0,9 | potassium | 19870,0 | -0,2 |
| copper | 0,2 | -0,5 | cadmium | 0,0 | -0,3 |
| germanium | 0,0 | -0,5 | aluminium | 8,1 | -0,3 |
| iron | 8,7 | -0,4 | molybdenum | 0,0 | -0,4 |
| lead | 0,0 | -0,4 | calcium | 1426,0 | -0,4 |
| lithium | 0,4 | -0,2 | manganese | 8,3 | -0,4 |
| magnesium | 2657,0 | 0,2 | selenium | 0,0 | -0,4 |
| manganese | 8,3 | -0,4 | iron | 8,7 | -0,4 |
| mercury | 0,0 | -0,4 | lead | 0,0 | -0,4 |
| molybdenum | 0,0 | -0,4 | mercury | 0,0 | -0,4 |
| nickel | 0,0 | -0,6 | copper | 0,2 | -0,5 |
| phosphorus | 588,8 | -0,7 | germanium | 0,0 | -0,5 |
| potassium | 19870,0 | -0,2 | nickel | 0,0 | -0,6 |
| selenium | 0,0 | -0,4 | phosphorus | 588,8 | -0,7 |
| sodium | 21220,0 | 5,7 | vanadium | 0,0 | -0,7 |
| vanadium | 0,0 | -0,7 | zinc | 3,8 | -0,7 |
| zinc | 3,8 | -0,7 | cobalt | 0,0 | -0,9 |

# Fumaria officinalis

| | |
|---|---|
| Family: | Fumariaceae |
| Scientific name: | Fumaria officinalis |
| English name: | Fumitory |
| German name: | |
| French name: | Fumeterre |
| Dutch name: | Duivekervel |
| Part Used: | herba |
| Extraction: | watery |
| Titer: | 1/5 |
| Standard minimal: | 1% fumarin |
| Comment: | |

| Sorted by element name | | | Sorted by relative difference | | |
|---|---|---|---|---|---|
| aluminium | 78,0 | 0,7 | calcium | 15100,0 | 2,4 |
| arsenicum | 0,0 | -0,4 | potassium | 50640,0 | 1,1 |
| cadmium | 0,3 | 0,7 | aluminium | 78,0 | 0,7 |
| calcium | 15100,0 | 2,4 | cadmium | 0,3 | 0,7 |
| chromium | 0,4 | 0,3 | zinc | 24,4 | 0,5 |
| cobalt | 0,3 | -0,1 | chromium | 0,4 | 0,3 |
| copper | 4,6 | -0,1 | iron | 83,4 | 0,1 |
| germanium | 0,0 | -0,5 | manganese | 57,8 | 0,0 |
| iron | 83,4 | 0,1 | cobalt | 0,3 | -0,1 |
| lead | 0,0 | -0,4 | copper | 4,6 | -0,1 |
| lithium | 0,0 | -0,5 | magnesium | 1518,0 | -0,3 |
| magnesium | 1518,0 | -0,3 | molybdenum | 0,0 | -0,4 |
| manganese | 57,8 | 0,0 | phosphorus | 976,0 | -0,4 |
| mercury | 0,0 | -0,4 | selenium | 0,0 | -0,4 |
| molybdenum | 0,0 | -0,4 | vanadium | 0,2 | -0,4 |
| nickel | 0,4 | -0,5 | sodium | 570,3 | -0,4 |
| phosphorus | 976,0 | -0,4 | arsenicum | 0,0 | -0,4 |
| potassium | 50640,0 | 1,1 | lead | 0,0 | -0,4 |
| selenium | 0,0 | -0,4 | mercury | 0,0 | -0,4 |
| sodium | 570,3 | -0,4 | lithium | 0,0 | -0,5 |
| vanadium | 0,2 | -0,4 | germanium | 0,0 | -0,5 |
| zinc | 24,4 | 0,5 | nickel | 0,4 | -0,5 |

# Gentiana lutea

| | |
|---|---|
| Family: | Gentianaceae |
| Scientific name: | Gentiana lutea |
| English name: | Gentian yellow |
| German name: | Gelber Enzian, Bitterwurz |
| French name: | Gentiane jaune |
| Dutch name: | Gele gentiaan |
| Part Used: | radix |
| Extraction: | watery |
| Titer: | 1/4 |
| Standard minimal: | |
| Comment: | |

| Sorted by element name | | | Sorted by relative difference | | |
|---|---|---|---|---|---|
| aluminium | 110,7 | 1,2 | cobalt | 1,5 | 3,0 |
| arsenicum | 0,0 | -0,4 | germanium | 2,6 | 2,9 |
| cadmium | 0,0 | -0,3 | aluminium | 110,7 | 1,2 |
| calcium | 1568,0 | -0,3 | nickel | 3,3 | 0,2 |
| chromium | 0,0 | -0,6 | sodium | 2557,0 | 0,2 |
| cobalt | 1,5 | 3,0 | iron | 74,4 | 0,0 |
| copper | 2,6 | -0,3 | manganese | 49,0 | -0,1 |
| germanium | 2,6 | 2,9 | cadmium | 0,0 | -0,3 |
| iron | 74,4 | 0,0 | copper | 2,6 | -0,3 |
| lead | 0,0 | -0,4 | magnesium | 1594,0 | -0,3 |
| lithium | 0,0 | -0,5 | calcium | 1568,0 | -0,3 |
| magnesium | 1594,0 | -0,3 | molybdenum | 0,0 | -0,4 |
| manganese | 49,0 | -0,1 | selenium | 0,0 | -0,4 |
| mercury | 0,0 | -0,4 | zinc | 9,7 | -0,4 |
| molybdenum | 0,0 | -0,4 | arsenicum | 0,0 | -0,4 |
| nickel | 3,3 | 0,2 | lead | 0,0 | -0,4 |
| phosphorus | 672,0 | -0,6 | mercury | 0,0 | -0,4 |
| potassium | 5587,0 | -0,8 | lithium | 0,0 | -0,5 |
| selenium | 0,0 | -0,4 | chromium | 0,0 | -0,6 |
| sodium | 2557,0 | 0,2 | phosphorus | 672,0 | -0,6 |
| vanadium | 0,0 | -0,7 | vanadium | 0,0 | -0,7 |
| zinc | 9,7 | -0,4 | potassium | 5587,0 | -0,8 |

# Ginkgo biloba

| | |
|---|---|
| Family: | Ginkgoaceae |
| Scientific name: | Ginkgo biloba |
| English name: | Maidentree, Fossiltree |
| German name: | Ginkgobaum |
| French name: | Noyer du japon |
| Dutch name: | Ginkgo |
| Part Used: | folia |
| Extraction: | alcoholic watery |
| Titer: | 1/6 |
| Standard minimal: | 1% flavonoids |
| Comment: | |

| Sorted by element name | | | Sorted by relative difference | | |
|---|---|---|---|---|---|
| aluminium | 13,4 | -0,3 | magnesium | 7326,0 | 2,6 |
| arsenicum | 0,7 | -0,1 | potassium | 55360,0 | 1,3 |
| cadmium | 0,0 | -0,3 | lithium | 2,7 | 1,2 |
| calcium | 4043,0 | 0,2 | phosphorus | 3241,0 | 1,2 |
| chromium | 0,2 | -0,2 | calcium | 4043,0 | 0,2 |
| cobalt | 0,2 | -0,3 | arsenicum | 0,7 | -0,1 |
| copper | 1,5 | -0,4 | chromium | 0,2 | -0,2 |
| germanium | 0,0 | -0,5 | manganese | 39,2 | -0,2 |
| iron | 37,7 | -0,2 | iron | 37,7 | -0,2 |
| lead | 0,0 | -0,4 | aluminium | 13,4 | -0,3 |
| lithium | 2,7 | 1,2 | cadmium | 0,0 | -0,3 |
| magnesium | 7326,0 | 2,6 | sodium | 1012,0 | -0,3 |
| manganese | 39,2 | -0,2 | cobalt | 0,2 | -0,3 |
| mercury | 0,0 | -0,4 | copper | 1,5 | -0,4 |
| molybdenum | 0,0 | -0,4 | molybdenum | 0,0 | -0,4 |
| nickel | 0,9 | -0,4 | selenium | 0,0 | -0,4 |
| phosphorus | 3241,0 | 1,2 | nickel | 0,9 | -0,4 |
| potassium | 55360,0 | 1,3 | lead | 0,0 | -0,4 |
| selenium | 0,0 | -0,4 | mercury | 0,0 | -0,4 |
| sodium | 1012,0 | -0,3 | germanium | 0,0 | -0,5 |
| vanadium | 0,1 | -0,6 | zinc | 7,5 | -0,5 |
| zinc | 7,5 | -0,5 | vanadium | 0,1 | -0,6 |

# Ginseng

| | |
|---|---|
| Family: | Araliaceae |
| Scientific name: | Panax ginseng |
| English name: | Ginseng |
| German name: | Ginseng, Kraftwurzel |
| French name: | Ginseng |
| Dutch name: | Ginseng |
| Part Used: | radix |
| Extraction: | alcoholic watery |
| Titer: | 1/1,5 |
| Standard minimal: | 1% ginsenosid |
| Comment: | |

| Sorted by element name | | | Sorted by relative difference | | |
|---|---|---|---|---|---|
| aluminium | 2,9 | -0,4 | germanium | 1,6 | 1,6 |
| arsenicum | 0,0 | -0,4 | cobalt | 0,9 | 1,5 |
| cadmium | 0,1 | 0,1 | vanadium | 0,7 | 0,3 |
| calcium | 178,2 | -0,6 | sodium | 2676,0 | 0,2 |
| chromium | 0,0 | -0,6 | cadmium | 0,1 | 0,1 |
| cobalt | 0,9 | 1,5 | copper | 4,1 | -0,2 |
| copper | 4,1 | -0,2 | molybdenum | 0,0 | -0,4 |
| germanium | 1,6 | 1,6 | selenium | 0,0 | -0,4 |
| iron | 2,9 | -0,5 | manganese | 3,2 | -0,4 |
| lead | 0,0 | -0,4 | aluminium | 2,9 | -0,4 |
| lithium | 0,0 | -0,5 | arsenicum | 0,0 | -0,4 |
| magnesium | 250,9 | -1,0 | lead | 0,0 | -0,4 |
| manganese | 3,2 | -0,4 | mercury | 0,0 | -0,4 |
| mercury | 0,0 | -0,4 | iron | 2,9 | -0,5 |
| molybdenum | 0,0 | -0,4 | lithium | 0,0 | -0,5 |
| nickel | 0,7 | -0,5 | nickel | 0,7 | -0,5 |
| phosphorus | 366,7 | -0,8 | chromium | 0,0 | -0,6 |
| potassium | 3062,0 | -0,9 | calcium | 178,2 | -0,6 |
| selenium | 0,0 | -0,4 | zinc | 5,4 | -0,6 |
| sodium | 2676,0 | 0,2 | phosphorus | 366,7 | -0,8 |
| vanadium | 0,7 | 0,3 | potassium | 3062,0 | -0,9 |
| zinc | 5,4 | -0,6 | magnesium | 250,9 | -1,0 |

# Glycyrrhiza glabra

| | |
|---|---|
| Family: | Leguminosae |
| Scientific name: | Glycyrrhiza glabra |
| English name: | Liquorice |
| German name: | Lakritze |
| French name: | Réglisse officinale |
| Dutch name: | Zoethout |
| Part Used: | radix |
| Extraction: | watery |
| Titer: | 1/4 |
| Standard minimal: | 10% glycyrrhizin acid |
| Comment: | |

| Sorted by element name | | | Sorted by relative difference | | |
|---|---|---|---|---|---|
| aluminium | 19,0 | -0,2 | molybdenum | 1,8 | 2,3 |
| arsenicum | 0,4 | -0,2 | cobalt | 1,0 | 1,7 |
| cadmium | 0,0 | -0,3 | germanium | 1,6 | 1,6 |
| calcium | 3824,0 | 0,1 | sodium | 6973,0 | 1,5 |
| chromium | 0,2 | -0,2 | vanadium | 1,3 | 1,2 |
| cobalt | 1,0 | 1,7 | magnesium | 4076,0 | 1,0 |
| copper | 9,4 | 0,2 | iron | 181,6 | 0,8 |
| germanium | 1,6 | 1,6 | lithium | 1,3 | 0,3 |
| iron | 181,6 | 0,8 | copper | 9,4 | 0,2 |
| lead | 0,0 | -0,4 | calcium | 3824,0 | 0,1 |
| lithium | 1,3 | 0,3 | potassium | 22950,0 | -0,1 |
| magnesium | 4076,0 | 1,0 | chromium | 0,2 | -0,2 |
| manganese | 14,4 | -0,3 | aluminium | 19,0 | -0,2 |
| mercury | 0,0 | -0,4 | arsenicum | 0,4 | -0,2 |
| molybdenum | 1,8 | 2,3 | cadmium | 0,0 | -0,3 |
| nickel | 0,5 | -0,5 | zinc | 10,9 | -0,3 |
| phosphorus | 463,0 | -0,7 | manganese | 14,4 | -0,3 |
| potassium | 22950,0 | -0,1 | selenium | 0,0 | -0,4 |
| selenium | 0,0 | -0,4 | lead | 0,0 | -0,4 |
| sodium | 6973,0 | 1,5 | mercury | 0,0 | -0,4 |
| vanadium | 1,3 | 1,2 | nickel | 0,5 | -0,5 |
| zinc | 10,9 | -0,3 | phosphorus | 463,0 | -0,7 |

# Hamamelis virginia

| | |
|---|---|
| Family: | Hamamelidae |
| Scientific name: | Hamamelis virginiana |
| English name: | Witch hazel |
| German name: | Virginischer Zauberstrauch |
| French name: | Noisetier de la sorcière |
| Dutch name: | Toverhazelaar |
| Part Used: | folia |
| Extraction: | alcoholic watery |
| Titer: | 1/4 |
| Standard minimal: | 23% hamamelis tannin |
| Comment: | |

| Sorted by element name | | | Sorted by relative difference | | |
|---|---|---|---|---|---|
| aluminium | 2,5 | -0,4 | lead | 1,5 | 1,1 |
| arsenicum | 0,0 | -0,4 | manganese | 141,0 | 0,5 |
| cadmium | 0,0 | -0,3 | germanium | 0,6 | 0,3 |
| calcium | 1306,0 | -0,4 | vanadium | 0,6 | 0,2 |
| chromium | 0,0 | -0,6 | cadmium | 0,0 | -0,3 |
| cobalt | 0,0 | -0,8 | copper | 1,3 | -0,4 |
| copper | 1,3 | -0,4 | molybdenum | 0,0 | -0,4 |
| germanium | 0,6 | 0,3 | calcium | 1306,0 | -0,4 |
| iron | 12,1 | -0,4 | iron | 12,1 | -0,4 |
| lead | 1,5 | 1,1 | selenium | 0,0 | -0,4 |
| lithium | 0,1 | -0,4 | aluminium | 2,5 | -0,4 |
| magnesium | 1059,0 | -0,6 | lithium | 0,1 | -0,4 |
| manganese | 141,0 | 0,5 | arsenicum | 0,0 | -0,4 |
| mercury | 0,0 | -0,4 | mercury | 0,0 | -0,4 |
| molybdenum | 0,0 | -0,4 | sodium | 401,6 | -0,5 |
| nickel | 0,0 | -0,6 | chromium | 0,0 | -0,6 |
| phosphorus | 593,1 | -0,7 | magnesium | 1059,0 | -0,6 |
| potassium | 8374,0 | -0,7 | nickel | 0,0 | -0,6 |
| selenium | 0,0 | -0,4 | zinc | 5,3 | -0,6 |
| sodium | 401,6 | -0,5 | phosphorus | 593,1 | -0,7 |
| vanadium | 0,6 | 0,2 | potassium | 8374,0 | -0,7 |
| zinc | 5,3 | -0,6 | cobalt | 0,0 | -0,8 |

# Harpagophytum procumbens

| | |
|---|---|
| Family: | Pedaliaceae |
| Scientific name: | Harpagophytum procumbens |
| English name: | Woodspider |
| German name: | Teufelskralle |
| French name: | Griffe du diable |
| Dutch name: | Duivelsklauw, Wolspin |
| Part Used: | radix |
| Extraction: | alcoholic watery |
| Titer: | 1/2 |
| Standard minimal: | 1% harpagosid |
| Comment: | |

| Sorted by element name | | | Sorted by relative difference | | |
|---|---|---|---|---|---|
| aluminium | 10,4 | -0,3 | germanium | 0,4 | 0,0 |
| arsenicum | 0,0 | -0,4 | chromium | 0,2 | -0,2 |
| cadmium | 0,0 | -0,3 | sodium | 1227,0 | -0,2 |
| calcium | 436,5 | -0,6 | cadmium | 0,0 | -0,3 |
| chromium | 0,2 | -0,2 | aluminium | 10,4 | -0,3 |
| cobalt | 0,0 | -0,8 | molybdenum | 0,0 | -0,4 |
| copper | 0,6 | -0,4 | manganese | 8,7 | -0,4 |
| germanium | 0,4 | 0,0 | selenium | 0,0 | -0,4 |
| iron | 8,0 | -0,4 | copper | 0,6 | -0,4 |
| lead | 0,0 | -0,4 | iron | 8,0 | -0,4 |
| lithium | 0,0 | -0,5 | arsenicum | 0,0 | -0,4 |
| magnesium | 1057,0 | -0,6 | lead | 0,0 | -0,4 |
| manganese | 8,7 | -0,4 | mercury | 0,0 | -0,4 |
| mercury | 0,0 | -0,4 | lithium | 0,0 | -0,5 |
| molybdenum | 0,0 | -0,4 | calcium | 436,5 | -0,6 |
| nickel | 0,1 | -0,6 | magnesium | 1057,0 | -0,6 |
| phosphorus | 207,9 | -0,9 | nickel | 0,1 | -0,6 |
| potassium | 5785,0 | -0,8 | vanadium | 0,0 | -0,7 |
| selenium | 0,0 | -0,4 | zinc | 3,4 | -0,8 |
| sodium | 1227,0 | -0,2 | cobalt | 0,0 | -0,8 |
| vanadium | 0,0 | -0,7 | potassium | 5785,0 | -0,8 |
| zinc | 3,4 | -0,8 | phosphorus | 207,9 | -0,9 |

# Hibiscus sabdariffa

Family:               Malvaceae
Scientific name:      Hibiscus sabdariffa
English name:
German name:          Rote sauerampferappel
French name:          Bisap du Senegal
Dutch name:           Hibiscus
Part Used:            flores
Extraction:           watery
Titer:                1/4
Standard minimal:
Comment:

| Sorted by element name | | | Sorted by relative difference | | |
|---|---|---|---|---|---|
| aluminium | 51,3 | 0,3 | manganese | 864,4 | 5,4 |
| arsenicum | 0,9 | 0,0 | cobalt | 1,4 | 2,7 |
| cadmium | 0,0 | -0,3 | vanadium | 2,1 | 2,4 |
| calcium | 7396,0 | 0,8 | magnesium | 4500,0 | 1,2 |
| chromium | 0,6 | 0,7 | zinc | 31,6 | 0,9 |
| cobalt | 1,4 | 2,7 | calcium | 7396,0 | 0,8 |
| copper | 2,7 | -0,3 | chromium | 0,6 | 0,7 |
| germanium | 0,0 | -0,5 | potassium | 40110,0 | 0,6 |
| iron | 95,3 | 0,2 | nickel | 4,8 | 0,6 |
| lead | 0,0 | -0,4 | aluminium | 51,3 | 0,3 |
| lithium | 0,3 | -0,3 | iron | 95,3 | 0,2 |
| magnesium | 4500,0 | 1,2 | arsenicum | 0,9 | 0,0 |
| manganese | 864,4 | 5,4 | cadmium | 0,0 | -0,3 |
| mercury | 0,0 | -0,4 | copper | 2,7 | -0,3 |
| molybdenum | 0,0 | -0,4 | lithium | 0,3 | -0,3 |
| nickel | 4,8 | 0,6 | molybdenum | 0,0 | -0,4 |
| phosphorus | 951,4 | -0,4 | selenium | 0,0 | -0,4 |
| potassium | 40110,0 | 0,6 | phosphorus | 951,4 | -0,4 |
| selenium | 0,0 | -0,4 | lead | 0,0 | -0,4 |
| sodium | 497,8 | -0,5 | mercury | 0,0 | -0,4 |
| vanadium | 2,1 | 2,4 | sodium | 497,8 | -0,5 |
| zinc | 31,6 | 0,9 | germanium | 0,0 | -0,5 |

# Hieracium pilosella

Family:              Compositae
Scientific name:     Hieracium pilosella
English name:        Hawkweed, Mouse-ear
German name:         Behaartes Habichtskraut
French name:         Piloselle, Oreille de souris
Dutch name:          Muizeoor
Part Used:           herba
Extraction:          alcoholic
Titer:               1/6
Standard minimal:    0,5% umbelliferon
Comment:

| Sorted by element name | | | Sorted by relative difference | | |
|---|---|---|---|---|---|
| aluminium | 296,5 | 4,1 | mercury | 2,4 | 4,4 |
| arsenicum | 7,8 | 3,5 | aluminium | 296,5 | 4,1 |
| cadmium | 0,0 | -0,3 | iron | 651,9 | 4,0 |
| calcium | 3900,0 | 0,1 | arsenicum | 7,8 | 3,5 |
| chromium | 0,5 | 0,5 | lead | 1,9 | 1,5 |
| cobalt | 0,5 | 0,4 | potassium | 53870,0 | 1,2 |
| copper | 16,5 | 0,7 | vanadium | 1,3 | 1,2 |
| germanium | 0,0 | -0,5 | zinc | 31,5 | 0,9 |
| iron | 651,9 | 4,0 | phosphorus | 2709,0 | 0,8 |
| lead | 1,9 | 1,5 | copper | 16,5 | 0,7 |
| lithium | 0,3 | -0,3 | chromium | 0,5 | 0,5 |
| magnesium | 2060,0 | -0,1 | cobalt | 0,5 | 0,4 |
| manganese | 81,3 | 0,1 | calcium | 3900,0 | 0,1 |
| mercury | 2,4 | 4,4 | manganese | 81,3 | 0,1 |
| molybdenum | 0,0 | -0,4 | magnesium | 2060,0 | -0,1 |
| nickel | 2,1 | -0,1 | nickel | 2,1 | -0,1 |
| phosphorus | 2709,0 | 0,8 | cadmium | 0,0 | -0,3 |
| potassium | 53870,0 | 1,2 | sodium | 1144,0 | -0,3 |
| selenium | 0,0 | -0,4 | lithium | 0,3 | -0,3 |
| sodium | 1144,0 | -0,3 | molybdenum | 0,0 | -0,4 |
| vanadium | 1,3 | 1,2 | selenium | 0,0 | -0,4 |
| zinc | 31,5 | 0,9 | germanium | 0,0 | -0,5 |

# Humulus lupulus

| | |
|---|---|
| Family: | Moraceae |
| Scientific name: | Humulus lupulus |
| English name: | Hops |
| German name: | Hopfen |
| French name: | Houblon |
| Dutch name: | Hop |
| Part Used: | flores strobuli |
| Extraction: | alcoholic watery |
| Titer: | 1/6 |
| Standard minimal: | 0,4% flavonoids |
| Comment: | |

| Sorted by element name | | | Sorted by relative difference | | |
|---|---|---|---|---|---|
| aluminium | 1,1 | -0,4 | copper | 63,9 | 4,2 |
| arsenicum | 0,3 | -0,3 | selenium | 6,5 | 1,6 |
| cadmium | 0,0 | -0,3 | potassium | 26470,0 | 0,1 |
| calcium | 71,1 | -0,6 | lithium | 0,4 | -0,2 |
| chromium | 0,0 | -0,5 | cadmium | 0,0 | -0,3 |
| cobalt | 0,0 | -0,9 | arsenicum | 0,3 | -0,3 |
| copper | 63,9 | 4,2 | molybdenum | 0,0 | -0,3 |
| germanium | 0,0 | -0,5 | manganese | 0,2 | -0,4 |
| iron | 5,4 | -0,4 | aluminium | 1,1 | -0,4 |
| lead | 0,0 | -0,4 | iron | 5,4 | -0,4 |
| lithium | 0,4 | -0,2 | lead | 0,0 | -0,4 |
| magnesium | 47,7 | -1,1 | mercury | 0,0 | -0,4 |
| manganese | 0,2 | -0,4 | germanium | 0,0 | -0,5 |
| mercury | 0,0 | -0,4 | chromium | 0,0 | -0,5 |
| molybdenum | 0,0 | -0,3 | nickel | 0,3 | -0,6 |
| nickel | 0,3 | -0,6 | sodium | 125,5 | -0,6 |
| phosphorus | 622,6 | -0,6 | calcium | 71,1 | -0,6 |
| potassium | 26470,0 | 0,1 | phosphorus | 622,6 | -0,6 |
| selenium | 6,5 | 1,6 | vanadium | 0,0 | -0,7 |
| sodium | 125,5 | -0,6 | cobalt | 0,0 | -0,9 |
| vanadium | 0,0 | -0,7 | zinc | 0,7 | -0,9 |
| zinc | 0,7 | -0,9 | magnesium | 47,7 | -1,1 |

# Hydrocotyle asiatica

| | |
|---|---|
| Family: | Umbelliferae |
| Scientific name: | Centella asiatica |
| English name: | Indian pennywort |
| German name: | Asiatische Wassernessel |
| French name: | Hydrocotyle asiatique |
| Dutch name: | Aziatische waternavel |
| Part Used: | totum saponique |
| Extraction: | |
| Titer: | 1/ |
| Standard minimal: | 9% asiatocid, 10% madecassosid |
| Comment: | |

| Sorted by element name | | | Sorted by relative difference | | |
|---|---|---|---|---|---|
| aluminium | 14,2 | -0,2 | cobalt | 1,4 | 2,7 |
| arsenicum | 0,0 | -0,4 | mercury | 1,0 | 1,6 |
| cadmium | 0,0 | -0,3 | sodium | 5774,0 | 1,1 |
| calcium | 596,5 | -0,5 | nickel | 4,0 | 0,4 |
| chromium | 0,1 | -0,5 | molybdenum | 0,5 | 0,4 |
| cobalt | 1,4 | 2,7 | manganese | 93,4 | 0,2 |
| copper | 5,9 | 0,0 | potassium | 29720,0 | 0,2 |
| germanium | 0,0 | -0,5 | lead | 0,6 | 0,2 |
| iron | 33,0 | -0,3 | zinc | 17,9 | 0,1 |
| lead | 0,6 | 0,2 | magnesium | 2227,0 | 0,0 |
| lithium | 0,1 | -0,4 | copper | 5,9 | 0,0 |
| magnesium | 2227,0 | 0,0 | aluminium | 14,2 | -0,2 |
| manganese | 93,4 | 0,2 | iron | 33,0 | -0,3 |
| mercury | 1,0 | 1,6 | cadmium | 0,0 | -0,3 |
| molybdenum | 0,5 | 0,4 | phosphorus | 1024,0 | -0,4 |
| nickel | 4,0 | 0,4 | selenium | 0,0 | -0,4 |
| phosphorus | 1024,0 | -0,4 | lithium | 0,1 | -0,4 |
| potassium | 29720,0 | 0,2 | arsenicum | 0,0 | -0,4 |
| selenium | 0,0 | -0,4 | chromium | 0,1 | -0,5 |
| sodium | 5774,0 | 1,1 | germanium | 0,0 | -0,5 |
| vanadium | 0,0 | -0,7 | calcium | 596,5 | -0,5 |
| zinc | 17,9 | 0,1 | vanadium | 0,0 | -0,7 |

# Hyoscyamus niger

Family:                  solanaceae
Scientific name:         hyoscyamus niger
English name:            Henbane
German name:             Bilsenkraut, Schweinekraut
French name:             Jusquiame
Dutch name:              Bilzekruid
Part Used:               herba
Extraction:              alcoholic watery
Titer:                   1/
Standard minimal:        0,4% alkaloids
Comment:                 the extreme high lithium levels
confirm the parallel in the picture of Lithium and
Hyoscyamus.

| Sorted by element name | | | Sorted by relative difference | | |
|---|---|---|---|---|---|
| aluminium | 12,5 | -0,3 | lithium | 11,9 | 6,8 |
| arsenicum | 0,0 | -0,4 | molybdenum | 1,4 | 1,7 |
| cadmium | 0,0 | -0,3 | potassium | 53000,0 | 1,2 |
| calcium | 352,8 | -0,6 | zinc | 34,4 | 1,0 |
| chromium | 0,1 | -0,4 | magnesium | 3063,0 | 0,5 |
| cobalt | 0,2 | -0,3 | copper | 12,3 | 0,4 |
| copper | 12,3 | 0,4 | germanium | 0,6 | 0,3 |
| germanium | 0,6 | 0,3 | sodium | 2800,0 | 0,2 |
| iron | 27,8 | -0,3 | cadmium | 0,0 | -0,3 |
| lead | 0,0 | -0,4 | aluminium | 12,5 | -0,3 |
| lithium | 11,9 | 6,8 | iron | 27,8 | -0,3 |
| magnesium | 3063,0 | 0,5 | cobalt | 0,2 | -0,3 |
| manganese | 15,6 | -0,3 | manganese | 15,6 | -0,3 |
| mercury | 0,0 | -0,4 | chromium | 0,1 | -0,4 |
| molybdenum | 1,4 | 1,7 | selenium | 0,0 | -0,4 |
| nickel | 0,3 | -0,6 | phosphorus | 965,7 | -0,4 |
| phosphorus | 965,7 | -0,4 | vanadium | 0,2 | -0,4 |
| potassium | 53000,0 | 1,2 | arsenicum | 0,0 | -0,4 |
| selenium | 0,0 | -0,4 | lead | 0,0 | -0,4 |
| sodium | 2800,0 | 0,2 | mercury | 0,0 | -0,4 |
| vanadium | 0,2 | -0,4 | nickel | 0,3 | -0,6 |
| zinc | 34,4 | 1,0 | calcium | 352,8 | -0,6 |

# Hypericum perforatum

| | |
|---|---|
| Family: | Guttiferae |
| Scientific name: | Hypericum perforatum |
| English name: | St Johns wort |
| German name: | Johanneskraut |
| French name: | Mille pertuis |
| Dutch name: | St Janskruid, Hertshooi |
| Part Used: | herba |
| Extraction: | alcoholic watery |
| Titer: | 1/6 |
| Standard minimal: | 0,05% hypericin |
| Comment: | |

| Sorted by element name | | | Sorted by relative difference | | |
|---|---|---|---|---|---|
| aluminium | 1,2 | -0,4 | germanium | 1,5 | 1,5 |
| arsenicum | 0,0 | -0,4 | mercury | 0,7 | 1,0 |
| cadmium | 0,2 | 0,4 | cadmium | 0,2 | 0,4 |
| calcium | 227,2 | -0,6 | chromium | 0,3 | 0,1 |
| chromium | 0,3 | 0,1 | sodium | 1614,0 | -0,1 |
| cobalt | 0,0 | -0,9 | copper | 3,1 | -0,2 |
| copper | 3,1 | -0,2 | molybdenum | 0,0 | -0,4 |
| germanium | 1,5 | 1,5 | manganese | 6,1 | -0,4 |
| iron | 3,1 | -0,5 | selenium | 0,0 | -0,4 |
| lead | 0,0 | -0,4 | aluminium | 1,2 | -0,4 |
| lithium | 0,0 | -0,5 | arsenicum | 0,0 | -0,4 |
| magnesium | 237,2 | -1,0 | lead | 0,0 | -0,4 |
| manganese | 6,1 | -0,4 | iron | 3,1 | -0,5 |
| mercury | 0,7 | 1,0 | lithium | 0,0 | -0,5 |
| molybdenum | 0,0 | -0,4 | nickel | 0,6 | -0,5 |
| nickel | 0,6 | -0,5 | zinc | 6,7 | -0,6 |
| phosphorus | 584,1 | -0,7 | calcium | 227,2 | -0,6 |
| potassium | 4897,0 | -0,9 | phosphorus | 584,1 | -0,7 |
| selenium | 0,0 | -0,4 | vanadium | 0,0 | -0,7 |
| sodium | 1614,0 | -0,1 | cobalt | 0,0 | -0,9 |
| vanadium | 0,0 | -0,7 | potassium | 4897,0 | -0,9 |
| zinc | 6,7 | -0,6 | magnesium | 237,2 | -1,0 |

# Juniperus communis

| | |
|---|---|
| Family: | Cupressaceae |
| Scientific name: | Juniperus communis |
| English name: | Common juniper |
| German name: | Gemeiner Wacholder |
| French name: | Genévrier commun |
| Dutch name: | Jeneverbes |
| Part Used: | fructus |
| Extraction: | watery |
| Titer: | 1/2 |
| Standard minimal: | |
| Comment: | |

| Sorted by element name | | | Sorted by relative difference | | |
|---|---|---|---|---|---|
| aluminium | 4,7 | -0,4 | cobalt | 0,9 | 1,5 |
| arsenicum | 0,0 | -0,4 | lithium | 1,4 | 0,4 |
| cadmium | 0,0 | -0,3 | nickel | 2,9 | 0,1 |
| calcium | 656,7 | -0,5 | vanadium | 0,4 | -0,1 |
| chromium | 0,0 | -0,5 | manganese | 32,3 | -0,2 |
| cobalt | 0,9 | 1,5 | cadmium | 0,0 | -0,3 |
| copper | 1,5 | -0,4 | iron | 22,1 | -0,3 |
| germanium | 0,0 | -0,5 | copper | 1,5 | -0,4 |
| iron | 22,1 | -0,3 | molybdenum | 0,0 | -0,4 |
| lead | 0,0 | -0,4 | aluminium | 4,7 | -0,4 |
| lithium | 1,4 | 0,4 | selenium | 0,0 | -0,4 |
| magnesium | 464,3 | -0,9 | sodium | 696,7 | -0,4 |
| manganese | 32,3 | -0,2 | arsenicum | 0,0 | -0,4 |
| mercury | 0,0 | -0,4 | lead | 0,0 | -0,4 |
| molybdenum | 0,0 | -0,4 | mercury | 0,0 | -0,4 |
| nickel | 2,9 | 0,1 | germanium | 0,0 | -0,5 |
| phosphorus | 397,3 | -0,8 | calcium | 656,7 | -0,5 |
| potassium | 9631,0 | -0,7 | chromium | 0,0 | -0,5 |
| selenium | 0,0 | -0,4 | potassium | 9631,0 | -0,7 |
| sodium | 696,7 | -0,4 | zinc | 4,8 | -0,7 |
| vanadium | 0,4 | -0,1 | phosphorus | 397,3 | -0,8 |
| zinc | 4,8 | -0,7 | magnesium | 464,3 | -0,9 |

# Lespedeza capitata

Family:              Leguminosae
Scientific name:     Lespedeza capitata
English name:
German name:
French name:
Dutch name:          Lespedeza
Part Used:           herba
Extraction:          alcoholic watery
Titer:               1/5
Standard minimal:
Comment:

| Sorted by element name | | | Sorted by relative difference | | |
|---|---|---|---|---|---|
| aluminium | 8,9 | -0,3 | nickel | 4,4 | 0,5 |
| arsenicum | 0,5 | -0,2 | cobalt | 0,5 | 0,4 |
| cadmium | 0,2 | 0,3 | molybdenum | 0,5 | 0,4 |
| calcium | 615,0 | -0,5 | cadmium | 0,2 | 0,3 |
| chromium | 0,4 | 0,2 | zinc | 20,1 | 0,2 |
| cobalt | 0,5 | 0,4 | chromium | 0,4 | 0,2 |
| copper | 3,7 | -0,2 | arsenicum | 0,5 | -0,2 |
| germanium | 0,0 | -0,5 | copper | 3,7 | -0,2 |
| iron | 14,1 | -0,4 | manganese | 23,9 | -0,3 |
| lead | 0,0 | -0,4 | aluminium | 8,9 | -0,3 |
| lithium | 0,0 | -0,5 | sodium | 784,9 | -0,4 |
| magnesium | 1347,0 | -0,4 | iron | 14,1 | -0,4 |
| manganese | 23,9 | -0,3 | selenium | 0,0 | -0,4 |
| mercury | 0,0 | -0,4 | phosphorus | 965,5 | -0,4 |
| molybdenum | 0,5 | 0,4 | magnesium | 1347,0 | -0,4 |
| nickel | 4,4 | 0,5 | potassium | 14870,0 | -0,4 |
| phosphorus | 965,5 | -0,4 | lead | 0,0 | -0,4 |
| potassium | 14870,0 | -0,4 | mercury | 0,0 | -0,4 |
| selenium | 0,0 | -0,4 | lithium | 0,0 | -0,5 |
| sodium | 784,9 | -0,4 | germanium | 0,0 | -0,5 |
| vanadium | 0,0 | -0,7 | calcium | 615,0 | -0,5 |
| zinc | 20,1 | 0,2 | vanadium | 0,0 | -0,7 |

# Lithospermum

| | |
|---|---|
| Family: | Boraginaceae |
| Scientific name: | Lithospermum officinalis |
| English name: | Alkanet |
| German name: | Meerhjirse |
| French name: | Grémil |
| Dutch name: | Parelzaad, glad paerlkruid |
| Part Used: | herba |
| Extraction: | watery |
| Titer: | 1/5 |
| Standard minimal: | |
| Comment: | |

| Sorted by element name | | | Sorted by relative difference | | |
|---|---|---|---|---|---|
| aluminium | 18,3 | -0,2 | lead | 4,4 | 4,0 |
| arsenicum | 0,0 | -0,4 | potassium | 66620,0 | 1,7 |
| cadmium | 0,0 | -0,3 | calcium | 8548,0 | 1,1 |
| calcium | 8548,0 | 1,1 | zinc | 25,4 | 0,5 |
| chromium | 0,4 | 0,3 | vanadium | 0,8 | 0,5 |
| cobalt | 0,0 | -0,9 | chromium | 0,4 | 0,3 |
| copper | 3,7 | -0,2 | iron | 88,1 | 0,1 |
| germanium | 0,0 | -0,5 | phosphorus | 1517,0 | 0,0 |
| iron | 88,1 | 0,1 | lithium | 0,5 | -0,2 |
| lead | 4,4 | 4,0 | aluminium | 18,3 | -0,2 |
| lithium | 0,5 | -0,2 | magnesium | 1826,0 | -0,2 |
| magnesium | 1826,0 | -0,2 | copper | 3,7 | -0,2 |
| manganese | 29,0 | -0,2 | nickel | 1,6 | -0,2 |
| mercury | 0,0 | -0,4 | manganese | 29,0 | -0,2 |
| molybdenum | 0,0 | -0,4 | cadmium | 0,0 | -0,3 |
| nickel | 1,6 | -0,2 | sodium | 936,6 | -0,3 |
| phosphorus | 1517,0 | 0,0 | molybdenum | 0,0 | -0,4 |
| potassium | 66620,0 | 1,7 | selenium | 0,0 | -0,4 |
| selenium | 0,0 | -0,4 | arsenicum | 0,0 | -0,4 |
| sodium | 936,6 | -0,3 | mercury | 0,0 | -0,4 |
| vanadium | 0,8 | 0,5 | germanium | 0,0 | -0,5 |
| zinc | 25,4 | 0,5 | cobalt | 0,0 | -0,9 |

# Lotus corniculatus

| | |
|---|---|
| Family: | Nymphaeaceae |
| Scientific name: | Lotus corniculatus |
| English name: | Lotus |
| German name: | Lotus |
| French name: | Lotus |
| Dutch name: | Lotus |
| Part Used: | herba |
| Extraction: | watery |
| Titer: | 1/6 |
| Standard minimal: | |
| Comment: | |

| Sorted by element name | | | Sorted by relative difference | | |
|---|---|---|---|---|---|
| aluminium | 25,8 | -0,1 | phosphorus | 5506,0 | 2,7 |
| arsenicum | 2,8 | 1,0 | potassium | 84090,0 | 2,5 |
| cadmium | 0,3 | 0,7 | calcium | 13460,0 | 2,0 |
| calcium | 13460,0 | 2,0 | magnesium | 5990,0 | 1,9 |
| chromium | 0,6 | 0,7 | zinc | 46,0 | 1,7 |
| cobalt | 0,3 | -0,1 | arsenicum | 2,8 | 1,0 |
| copper | 14,8 | 0,6 | cadmium | 0,3 | 0,7 |
| germanium | 0,0 | -0,5 | chromium | 0,6 | 0,7 |
| iron | 96,5 | 0,2 | copper | 14,8 | 0,6 |
| lead | 0,1 | -0,4 | vanadium | 0,7 | 0,3 |
| lithium | 0,5 | -0,2 | iron | 96,5 | 0,2 |
| magnesium | 5990,0 | 1,9 | nickel | 3,1 | 0,2 |
| manganese | 60,5 | 0,0 | manganese | 60,5 | 0,0 |
| mercury | 0,0 | -0,4 | aluminium | 25,8 | -0,1 |
| molybdenum | 0,0 | -0,4 | cobalt | 0,3 | -0,1 |
| nickel | 3,1 | 0,2 | lithium | 0,5 | -0,2 |
| phosphorus | 5506,0 | 2,7 | sodium | 1289,0 | -0,2 |
| potassium | 84090,0 | 2,5 | molybdenum | 0,0 | -0,4 |
| selenium | 0,0 | -0,4 | lead | 0,1 | -0,4 |
| sodium | 1289,0 | -0,2 | selenium | 0,0 | -0,4 |
| vanadium | 0,7 | 0,3 | mercury | 0,0 | -0,4 |
| zinc | 46,0 | 1,7 | germanium | 0,0 | -0,5 |

# Marrubium vulgare

Family:             Labiatae
Scientific name:    Marrubium vulgare
English name:       Horehound
German name:        Andorin
French name:        Marrube
Dutch name:         Malrove, Andoorn
Part Used:          herba
Extraction:         watery
Titer:              1/5
Standard minimal:
Comment:

| Sorted by element name | | | Sorted by relative difference | | |
|---|---|---|---|---|---|
| aluminium | 12,0 | -0,3 | mercury | 0,8 | 1,2 |
| arsenicum | 0,0 | -0,4 | lead | 1,1 | 0,7 |
| cadmium | 0,0 | -0,3 | phosphorus | 2416,0 | 0,6 |
| calcium | 1365,0 | -0,4 | potassium | 36400,0 | 0,5 |
| chromium | 0,3 | 0,1 | cobalt | 0,5 | 0,4 |
| cobalt | 0,5 | 0,4 | molybdenum | 0,5 | 0,4 |
| copper | 2,6 | -0,3 | lithium | 1,1 | 0,2 |
| germanium | 0,0 | -0,5 | chromium | 0,3 | 0,1 |
| iron | 52,0 | -0,1 | magnesium | 2046,0 | -0,1 |
| lead | 1,1 | 0,7 | iron | 52,0 | -0,1 |
| lithium | 1,1 | 0,2 | sodium | 1438,0 | -0,2 |
| magnesium | 2046,0 | -0,1 | zinc | 13,5 | -0,2 |
| manganese | 15,1 | -0,3 | cadmium | 0,0 | -0,3 |
| mercury | 0,8 | 1,2 | copper | 2,6 | -0,3 |
| molybdenum | 0,5 | 0,4 | aluminium | 12,0 | -0,3 |
| nickel | 0,8 | -0,4 | manganese | 15,1 | -0,3 |
| phosphorus | 2416,0 | 0,6 | calcium | 1365,0 | -0,4 |
| potassium | 36400,0 | 0,5 | selenium | 0,0 | -0,4 |
| selenium | 0,0 | -0,4 | nickel | 0,8 | -0,4 |
| sodium | 1438,0 | -0,2 | arsenicum | 0,0 | -0,4 |
| vanadium | 0,0 | -0,7 | germanium | 0,0 | -0,5 |
| zinc | 13,5 | -0,2 | vanadium | 0,0 | -0,7 |

# Medicago sativa

| | |
|---|---|
| Family: | Leguminosae |
| Scientific name: | Medicago sativa |
| English name: | Alfalfa |
| German name: | Alfalfa, Luzerne |
| French name: | Luzerne cultivée |
| Dutch name: | Luzerne |
| Part Used: | herba |
| Extraction: | watery |
| Titer: | 1/6 |
| Standard minimal: | |
| Comment: | |

| Sorted by element name | | | Sorted by relative difference | | |
|---|---|---|---|---|---|
| aluminium | 6,9 | -0,4 | vanadium | 1,5 | 1,5 |
| arsenicum | 0,0 | -0,4 | potassium | 42780,0 | 0,7 |
| cadmium | 0,0 | -0,3 | mercury | 0,4 | 0,4 |
| calcium | 4222,0 | 0,2 | calcium | 4222,0 | 0,2 |
| chromium | 0,0 | -0,6 | lithium | 0,9 | 0,1 |
| cobalt | 0,0 | -0,9 | phosphorus | 1547,0 | 0,0 |
| copper | 0,9 | -0,4 | sodium | 1269,0 | -0,2 |
| germanium | 0,0 | -0,5 | cadmium | 0,0 | -0,3 |
| iron | 27,6 | -0,3 | iron | 27,6 | -0,3 |
| lead | 0,0 | -0,4 | magnesium | 1617,0 | -0,3 |
| lithium | 0,9 | 0,1 | manganese | 11,9 | -0,4 |
| magnesium | 1617,0 | -0,3 | aluminium | 6,9 | -0,4 |
| manganese | 11,9 | -0,4 | molybdenum | 0,0 | -0,4 |
| mercury | 0,4 | 0,4 | copper | 0,9 | -0,4 |
| molybdenum | 0,0 | -0,4 | selenium | 0,0 | -0,4 |
| nickel | 0,2 | -0,6 | zinc | 9,1 | -0,4 |
| phosphorus | 1547,0 | 0,0 | arsenicum | 0,0 | -0,4 |
| potassium | 42780,0 | 0,7 | lead | 0,0 | -0,4 |
| selenium | 0,0 | -0,4 | germanium | 0,0 | -0,5 |
| sodium | 1269,0 | -0,2 | chromium | 0,0 | -0,6 |
| vanadium | 1,5 | 1,5 | nickel | 0,2 | -0,6 |
| zinc | 9,1 | -0,4 | cobalt | 0,0 | -0,9 |

# Melilotus officinalis

| | |
|---|---|
| Family: | Leguminosae |
| Scientific name: | Melilotus officinalis |
| English name: | Sweet melilot |
| German name: | Honingklee |
| French name: | Mélilot, Couronne royale |
| Dutch name: | Honingklaver |
| Part Used: | herba |
| Extraction: | alcoholic watery |
| Titer: | 1/6,5 |
| Standard minimal: | 4% cumarin |
| Comment: | |

| Sorted by element name | | | Sorted by relative difference | | |
|---|---|---|---|---|---|
| aluminium | 3,3 | -0,4 | lead | 2,8 | 2,4 |
| arsenicum | 0,0 | -0,4 | vanadium | 1,3 | 1,2 |
| cadmium | 0,0 | -0,2 | magnesium | 3829,0 | 0,8 |
| calcium | 5353,0 | 0,4 | nickel | 5,0 | 0,6 |
| chromium | 0,2 | -0,2 | calcium | 5353,0 | 0,4 |
| cobalt | 0,0 | -0,9 | zinc | 23,2 | 0,4 |
| copper | 10,6 | 0,3 | potassium | 33630,0 | 0,4 |
| germanium | 0,0 | -0,5 | copper | 10,6 | 0,3 |
| iron | 12,2 | -0,4 | chromium | 0,2 | -0,2 |
| lead | 2,8 | 2,4 | cadmium | 0,0 | -0,2 |
| lithium | 0,4 | -0,2 | lithium | 0,4 | -0,2 |
| magnesium | 3829,0 | 0,8 | manganese | 17,0 | -0,3 |
| manganese | 17,0 | -0,3 | sodium | 823,3 | -0,4 |
| mercury | 0,0 | -0,4 | molybdenum | 0,0 | -0,4 |
| molybdenum | 0,0 | -0,4 | iron | 12,2 | -0,4 |
| nickel | 5,0 | 0,6 | selenium | 0,0 | -0,4 |
| phosphorus | 635,0 | -0,6 | aluminium | 3,3 | -0,4 |
| potassium | 33630,0 | 0,4 | arsenicum | 0,0 | -0,4 |
| selenium | 0,0 | -0,4 | mercury | 0,0 | -0,4 |
| sodium | 823,3 | -0,4 | germanium | 0,0 | -0,5 |
| vanadium | 1,3 | 1,2 | phosphorus | 635,0 | -0,6 |
| zinc | 23,2 | 0,4 | cobalt | 0,0 | -0,9 |

# Melissa officinalis

| | |
|---|---|
| Family: | Labiatae |
| Scientific name: | Melissa officinalis |
| English name: | lemon balm |
| German name: | Zitronenmelisse, Honingblume |
| French name: | Mélisse officinale |
| Dutch name: | Melisse, citroenmelisse |
| Part Used: | herba |
| Extraction: | watery |
| Titer: | 1/2,5 |
| Standard minimal: | 1,5% rosmarin acid |
| Comment: | |

| Sorted by element name | | | Sorted by relative difference | | |
|---|---|---|---|---|---|
| aluminium | 7,2 | -0,4 | vanadium | 0,7 | 0,3 |
| arsenicum | 0,0 | -0,4 | cadmium | 0,1 | 0,2 |
| cadmium | 0,1 | 0,2 | cobalt | 0,3 | -0,1 |
| calcium | 240,9 | -0,6 | copper | 2,6 | -0,3 |
| chromium | 0,0 | -0,6 | nickel | 1,2 | -0,3 |
| cobalt | 0,3 | -0,1 | aluminium | 7,2 | -0,4 |
| copper | 2,6 | -0,3 | molybdenum | 0,0 | -0,4 |
| germanium | 0,0 | -0,5 | selenium | 0,0 | -0,4 |
| iron | 4,0 | -0,5 | sodium | 682,0 | -0,4 |
| lead | 0,0 | -0,4 | manganese | 2,5 | -0,4 |
| lithium | 0,0 | -0,5 | arsenicum | 0,0 | -0,4 |
| magnesium | 305,9 | -1,0 | lead | 0,0 | -0,4 |
| manganese | 2,5 | -0,4 | mercury | 0,0 | -0,4 |
| mercury | 0,0 | -0,4 | iron | 4,0 | -0,5 |
| molybdenum | 0,0 | -0,4 | lithium | 0,0 | -0,5 |
| nickel | 1,2 | -0,3 | germanium | 0,0 | -0,5 |
| phosphorus | 215,1 | -0,9 | chromium | 0,0 | -0,6 |
| potassium | 3950,0 | -0,9 | calcium | 240,9 | -0,6 |
| selenium | 0,0 | -0,4 | zinc | 2,9 | -0,8 |
| sodium | 682,0 | -0,4 | potassium | 3950,0 | -0,9 |
| vanadium | 0,7 | 0,3 | phosphorus | 215,1 | -0,9 |
| zinc | 2,9 | -0,8 | magnesium | 305,9 | -1,0 |

# Mentha piperita

| | |
|---|---|
| Family: | Labiatae |
| Scientific name: | Mentha piperita |
| English name: | Peppermint |
| German name: | Pfefferminze |
| French name: | Menthe poivrée |
| Dutch name: | Pepermunt |
| Part Used: | folia |
| Extraction: | watery |
| Titer: | 1/4 |
| Standard minimal: | 3,5% etheric oil |
| Comment: | |

| Sorted by element name | | | Sorted by relative difference | | |
|---|---|---|---|---|---|
| aluminium | 6,6 | -0,4 | vanadium | 2,9 | 3,6 |
| arsenicum | 0,0 | -0,4 | magnesium | 3474,0 | 0,7 |
| cadmium | 0,1 | -0,1 | cobalt | 0,5 | 0,4 |
| calcium | 3896,0 | 0,1 | lithium | 1,3 | 0,3 |
| chromium | 0,0 | -0,6 | calcium | 3896,0 | 0,1 |
| cobalt | 0,5 | 0,4 | phosphorus | 1691,0 | 0,1 |
| copper | 0,6 | -0,4 | potassium | 27120,0 | 0,1 |
| germanium | 0,3 | -0,1 | sodium | 2247,0 | 0,1 |
| iron | 24,2 | -0,3 | manganese | 61,1 | 0,0 |
| lead | 0,0 | -0,4 | cadmium | 0,1 | -0,1 |
| lithium | 1,3 | 0,3 | germanium | 0,3 | -0,1 |
| magnesium | 3474,0 | 0,7 | iron | 24,2 | -0,3 |
| manganese | 61,1 | 0,0 | nickel | 1,3 | -0,3 |
| mercury | 0,0 | -0,4 | aluminium | 6,6 | -0,4 |
| molybdenum | 0,0 | -0,4 | molybdenum | 0,0 | -0,4 |
| nickel | 1,3 | -0,3 | selenium | 0,0 | -0,4 |
| phosphorus | 1691,0 | 0,1 | copper | 0,6 | -0,4 |
| potassium | 27120,0 | 0,1 | arsenicum | 0,0 | -0,4 |
| selenium | 0,0 | -0,4 | lead | 0,0 | -0,4 |
| sodium | 2247,0 | 0,1 | mercury | 0,0 | -0,4 |
| vanadium | 2,9 | 3,6 | chromium | 0,0 | -0,6 |
| zinc | 5,3 | -0,6 | zinc | 5,3 | -0,6 |

# Millefolium

| | |
|---|---|
| Family: | Compositae |
| Scientific name: | Achillea millefolium |
| English name: | Yarrow |
| German name: | Tausendblatt |
| French name: | Herbe du charpentier |
| Dutch name: | Duizendblad |
| Part Used: | herba |
| Extraction: | watery |
| Titer: | 1/6,5 |
| Standard minimal: | |
| Comment: | |

| Sorted by element name | | | Sorted by relative difference | | |
|---|---|---|---|---|---|
| aluminium | 0,0 | -0,5 | lithium | 2,3 | 1,0 |
| arsenicum | 0,0 | -0,4 | cadmium | 0,2 | 0,3 |
| cadmium | 0,2 | 0,3 | potassium | 23440,0 | -0,1 |
| calcium | 1378,0 | -0,4 | cobalt | 0,3 | -0,1 |
| chromium | 0,0 | -0,6 | phosphorus | 1331,0 | -0,2 |
| cobalt | 0,3 | -0,1 | sodium | 1321,0 | -0,2 |
| copper | 1,1 | -0,4 | nickel | 1,5 | -0,3 |
| germanium | 0,0 | -0,5 | manganese | 16,9 | -0,3 |
| iron | 17,1 | -0,4 | iron | 17,1 | -0,4 |
| lead | 0,0 | -0,4 | molybdenum | 0,0 | -0,4 |
| lithium | 2,3 | 1,0 | calcium | 1378,0 | -0,4 |
| magnesium | 1201,0 | -0,5 | copper | 1,1 | -0,4 |
| manganese | 16,9 | -0,3 | selenium | 0,0 | -0,4 |
| mercury | 0,0 | -0,4 | arsenicum | 0,0 | -0,4 |
| molybdenum | 0,0 | -0,4 | lead | 0,0 | -0,4 |
| nickel | 1,5 | -0,3 | mercury | 0,0 | -0,4 |
| phosphorus | 1331,0 | -0,2 | aluminium | 0,0 | -0,5 |
| potassium | 23440,0 | -0,1 | germanium | 0,0 | -0,5 |
| selenium | 0,0 | -0,4 | magnesium | 1201,0 | -0,5 |
| sodium | 1321,0 | -0,2 | zinc | 7,2 | -0,5 |
| vanadium | 0,0 | -0,7 | chromium | 0,0 | -0,6 |
| zinc | 7,2 | -0,5 | vanadium | 0,0 | -0,7 |

# Nasturtium officinalis

| | |
|---|---|
| Family: | Cruciferae |
| Scientific name: | Nasturtium officinalis |
| English name: | Watercresh |
| German name: | |
| French name: | Cresson de fontaine |
| Dutch name: | Waterkers witte |
| Part Used: | herba |
| Extraction: | watery |
| Titer: | 1/4 |
| Standard minimal: | |
| Comment: | |

| Sorted by element name | | | Sorted by relative difference | | |
|---|---|---|---|---|---|
| aluminium | 55,4 | 0,4 | potassium | 158400,0 | 5,6 |
| arsenicum | 8,0 | 3,6 | calcium | 28960,0 | 5,2 |
| cadmium | 0,0 | -0,3 | arsenicum | 8,0 | 3,6 |
| calcium | 28960,0 | 5,2 | vanadium | 1,9 | 2,1 |
| chromium | 0,8 | 1,1 | mercury | 1,2 | 2,0 |
| cobalt | 0,3 | -0,1 | sodium | 7629,0 | 1,7 |
| copper | 1,2 | -0,4 | magnesium | 5314,0 | 1,6 |
| germanium | 0,0 | -0,5 | phosphorus | 3821,0 | 1,6 |
| iron | 293,2 | 1,5 | iron | 293,2 | 1,5 |
| lead | 0,0 | -0,4 | chromium | 0,8 | 1,1 |
| lithium | 0,6 | -0,1 | zinc | 34,7 | 1,0 |
| magnesium | 5314,0 | 1,6 | aluminium | 55,4 | 0,4 |
| manganese | 70,9 | 0,0 | nickel | 3,4 | 0,2 |
| mercury | 1,2 | 2,0 | manganese | 70,9 | 0,0 |
| molybdenum | 0,0 | -0,4 | cobalt | 0,3 | -0,1 |
| nickel | 3,4 | 0,2 | lithium | 0,6 | -0,1 |
| phosphorus | 3821,0 | 1,6 | cadmium | 0,0 | -0,3 |
| potassium | 158400,0 | 5,6 | molybdenum | 0,0 | -0,4 |
| selenium | 0,0 | -0,4 | copper | 1,2 | -0,4 |
| sodium | 7629,0 | 1,7 | selenium | 0,0 | -0,4 |
| vanadium | 1,9 | 2,1 | lead | 0,0 | -0,4 |
| zinc | 34,7 | 1,0 | germanium | 0,0 | -0,5 |

# Olea europaea

| | |
|---|---|
| Family: | Oleaceae |
| Scientific name: | Olea europea |
| English name: | Olive tree |
| German name: | Olivebaum |
| French name: | Olivier |
| Dutch name: | Olijfboom |
| Part Used: | folia |
| Extraction: | alcoholic watery |
| Titer: | 1/5,5 |
| Standard minimal: | 3% oleuropein |
| Comment: | |

| Sorted by element name | | | Sorted by relative difference | | |
|---|---|---|---|---|---|
| aluminium | 5,6 | -0,4 | cobalt | 0,4 | 0,2 |
| arsenicum | 0,5 | -0,2 | lithium | 1,0 | 0,2 |
| cadmium | 0,0 | -0,3 | chromium | 0,3 | 0,1 |
| calcium | 1614,0 | -0,3 | arsenicum | 0,5 | -0,2 |
| chromium | 0,3 | 0,1 | copper | 3,5 | -0,2 |
| cobalt | 0,4 | 0,2 | germanium | 0,2 | -0,2 |
| copper | 3,5 | -0,2 | mercury | 0,1 | -0,2 |
| germanium | 0,2 | -0,2 | cadmium | 0,0 | -0,3 |
| iron | 13,3 | -0,4 | calcium | 1614,0 | -0,3 |
| lead | 0,0 | -0,4 | manganese | 13,3 | -0,3 |
| lithium | 1,0 | 0,2 | molybdenum | 0,0 | -0,4 |
| magnesium | 1297,0 | -0,5 | aluminium | 5,6 | -0,4 |
| manganese | 13,3 | -0,3 | iron | 13,3 | -0,4 |
| mercury | 0,1 | -0,2 | selenium | 0,0 | -0,4 |
| molybdenum | 0,0 | -0,4 | sodium | 673,3 | -0,4 |
| nickel | 0,8 | -0,4 | vanadium | 0,2 | -0,4 |
| phosphorus | 561,0 | -0,7 | nickel | 0,8 | -0,4 |
| potassium | 10040,0 | -0,6 | lead | 0,0 | -0,4 |
| selenium | 0,0 | -0,4 | magnesium | 1297,0 | -0,5 |
| sodium | 673,3 | -0,4 | potassium | 10040,0 | -0,6 |
| vanadium | 0,2 | -0,4 | phosphorus | 561,0 | -0,7 |
| zinc | 3,7 | -0,7 | zinc | 3,7 | -0,7 |

# Orthosiphon stamineus

| | |
|---|---|
| Family: | Labiatae |
| Scientific name: | Orthosiphon stamineus |
| English name: | |
| German name: | |
| French name: | |
| Dutch name: | Koemis koetjing, kattesnor |
| Part Used: | folia |
| Extraction: | alcoholic watery |
| Titer: | 1/4 |
| Standard minimal: | 0,1% sinensetin |
| Comment: | |

| Sorted by element name | | | Sorted by relative difference | | |
|---|---|---|---|---|---|
| aluminium | 2,9 | -0,4 | copper | 27,2 | 1,5 |
| arsenicum | 0,0 | -0,4 | cadmium | 0,3 | 0,6 |
| cadmium | 0,3 | 0,6 | zinc | 20,9 | 0,3 |
| calcium | 402,9 | -0,6 | molybdenum | 0,0 | -0,4 |
| chromium | 0,0 | -0,6 | iron | 14,3 | -0,4 |
| cobalt | 0,0 | -0,9 | selenium | 0,0 | -0,4 |
| copper | 27,2 | 1,5 | manganese | 4,7 | -0,4 |
| germanium | 0,0 | -0,5 | aluminium | 2,9 | -0,4 |
| iron | 14,3 | -0,4 | arsenicum | 0,0 | -0,4 |
| lead | 0,0 | -0,4 | lead | 0,0 | -0,4 |
| lithium | 0,0 | -0,5 | mercury | 0,0 | -0,4 |
| magnesium | 447,4 | -0,9 | lithium | 0,0 | -0,5 |
| manganese | 4,7 | -0,4 | germanium | 0,0 | -0,5 |
| mercury | 0,0 | -0,4 | sodium | 344,1 | -0,5 |
| molybdenum | 0,0 | -0,4 | nickel | 0,3 | -0,6 |
| nickel | 0,3 | -0,6 | calcium | 402,9 | -0,6 |
| phosphorus | 556,9 | -0,7 | chromium | 0,0 | -0,6 |
| potassium | 7040,0 | -0,8 | phosphorus | 556,9 | -0,7 |
| selenium | 0,0 | -0,4 | vanadium | 0,0 | -0,7 |
| sodium | 344,1 | -0,5 | potassium | 7040,0 | -0,8 |
| vanadium | 0,0 | -0,7 | cobalt | 0,0 | -0,9 |
| zinc | 20,9 | 0,3 | magnesium | 447,4 | -0,9 |

# Passiflora incarnata

| | |
|---|---|
| Family: | Passifloraceae |
| Scientific name: | Passiflora incarnata |
| English name: | Passion flower |
| German name: | Fleischfarbene Passionsblume |
| French name: | Fleur de Passion |
| Dutch name: | Passiebloem |
| Part Used: | herba |
| Extraction: | alcoholic watery |
| Titer: | 1/4 |
| Standard minimal: | 1,5% flavonoids |
| Comment: | |

| Sorted by element name | | | Sorted by relative difference | | |
|---|---|---|---|---|---|
| aluminium | 8,4 | -0,3 | zinc | 49,9 | 1,9 |
| arsenicum | 2,0 | 0,6 | lead | 1,7 | 1,3 |
| cadmium | 0,0 | -0,3 | arsenicum | 2,0 | 0,6 |
| calcium | 3173,0 | 0,0 | magnesium | 2976,0 | 0,4 |
| chromium | 0,0 | -0,6 | potassium | 32200,0 | 0,3 |
| cobalt | 0,0 | -0,9 | phosphorus | 1863,0 | 0,2 |
| copper | 2,7 | -0,3 | calcium | 3173,0 | 0,0 |
| germanium | 0,0 | -0,5 | nickel | 2,2 | -0,1 |
| iron | 27,1 | -0,3 | vanadium | 0,4 | -0,1 |
| lead | 1,7 | 1,3 | manganese | 35,5 | -0,2 |
| lithium | 0,4 | -0,2 | lithium | 0,4 | -0,2 |
| magnesium | 2976,0 | 0,4 | cadmium | 0,0 | -0,3 |
| manganese | 35,5 | -0,2 | copper | 2,7 | -0,3 |
| mercury | 0,0 | -0,4 | iron | 27,1 | -0,3 |
| molybdenum | 0,0 | -0,4 | sodium | 1040,0 | -0,3 |
| nickel | 2,2 | -0,1 | aluminium | 8,4 | -0,3 |
| phosphorus | 1863,0 | 0,2 | molybdenum | 0,0 | -0,4 |
| potassium | 32200,0 | 0,3 | selenium | 0,0 | -0,4 |
| selenium | 0,0 | -0,4 | mercury | 0,0 | -0,4 |
| sodium | 1040,0 | -0,3 | germanium | 0,0 | -0,5 |
| vanadium | 0,4 | -0,1 | chromium | 0,0 | -0,6 |
| zinc | 49,9 | 1,9 | cobalt | 0,0 | -0,9 |

# Phaseolus vulgaris

| | |
|---|---|
| Family: | Leguminosae |
| Scientific name: | Phaseolus vulgaris |
| English name: | French bean |
| German name: | Gartenbohne |
| French name: | Haricot à rames |
| Dutch name: | Boon gewone |
| Part Used: | fructus sine semen |
| Extraction: | alcoholic watery |
| Titer: | 1/10 |
| Standard minimal: | |
| Comment: | |

| Sorted by element name | | | Sorted by relative difference | | |
|---|---|---|---|---|---|
| aluminium | 4,5 | -0,4 | selenium | 13,3 | 3,7 |
| arsenicum | 0,0 | -0,4 | magnesium | 8270,0 | 3,1 |
| cadmium | 0,0 | -0,3 | lead | 3,5 | 3,1 |
| calcium | 1163,0 | -0,4 | potassium | 75040,0 | 2,1 |
| chromium | 0,2 | -0,2 | zinc | 50,4 | 1,9 |
| cobalt | 0,0 | -0,9 | molybdenum | 1,2 | 1,4 |
| copper | 22,8 | 1,2 | copper | 22,8 | 1,2 |
| germanium | 0,0 | -0,5 | nickel | 5,6 | 0,8 |
| iron | 33,4 | -0,2 | phosphorus | 2103,0 | 0,4 |
| lead | 3,5 | 3,1 | manganese | 48,2 | -0,1 |
| lithium | 0,4 | -0,2 | chromium | 0,2 | -0,2 |
| magnesium | 8270,0 | 3,1 | lithium | 0,4 | -0,2 |
| manganese | 48,2 | -0,1 | iron | 33,4 | -0,2 |
| mercury | 0,0 | -0,4 | cadmium | 0,0 | -0,3 |
| molybdenum | 1,2 | 1,4 | aluminium | 4,5 | -0,4 |
| nickel | 5,6 | 0,8 | calcium | 1163,0 | -0,4 |
| phosphorus | 2103,0 | 0,4 | arsenicum | 0,0 | -0,4 |
| potassium | 75040,0 | 2,1 | mercury | 0,0 | -0,4 |
| selenium | 13,3 | 3,7 | sodium | 454,2 | -0,5 |
| sodium | 454,2 | -0,5 | germanium | 0,0 | -0,5 |
| vanadium | 0,0 | -0,7 | vanadium | 0,0 | -0,7 |
| zinc | 50,4 | 1,9 | cobalt | 0,0 | -0,9 |

# Pinus maritima

| | |
|---|---|
| Family: | Pinaceae |
| Scientific name: | Pinus maritima |
| English name: | Pine |
| German name: | Kiefer, Föhre |
| French name: | Pin de mer |
| Dutch name: | Den |
| Part Used: | turniones gemmae |
| Extraction: | watery |
| Titer: | 1/5 |
| Standard minimal: | |
| Comment: | |

| Sorted by element name | | | Sorted by relative difference | | |
|---|---|---|---|---|---|
| aluminium | 166,4 | 2,1 | zinc | 97,2 | 4,6 |
| arsenicum | 1,1 | 0,1 | nickel | 17,0 | 3,7 |
| cadmium | 0,0 | -0,3 | aluminium | 166,4 | 2,1 |
| calcium | 3877,0 | 0,1 | manganese | 291,2 | 1,5 |
| chromium | 0,4 | 0,3 | selenium | 4,7 | 1,1 |
| cobalt | 0,3 | -0,1 | phosphorus | 3051,0 | 1,0 |
| copper | 8,7 | 0,2 | magnesium | 2766,0 | 0,3 |
| germanium | 0,0 | -0,5 | chromium | 0,4 | 0,3 |
| iron | 41,0 | -0,2 | lithium | 1,1 | 0,2 |
| lead | 0,0 | -0,4 | copper | 8,7 | 0,2 |
| lithium | 1,1 | 0,2 | arsenicum | 1,1 | 0,1 |
| magnesium | 2766,0 | 0,3 | calcium | 3877,0 | 0,1 |
| manganese | 291,2 | 1,5 | cobalt | 0,3 | -0,1 |
| mercury | 0,0 | -0,4 | iron | 41,0 | -0,2 |
| molybdenum | 0,0 | -0,4 | cadmium | 0,0 | -0,3 |
| nickel | 17,0 | 3,7 | molybdenum | 0,0 | -0,4 |
| phosphorus | 3051,0 | 1,0 | sodium | 778,9 | -0,4 |
| potassium | 10460,0 | -0,6 | lead | 0,0 | -0,4 |
| selenium | 4,7 | 1,1 | mercury | 0,0 | -0,4 |
| sodium | 778,9 | -0,4 | germanium | 0,0 | -0,5 |
| vanadium | 0,0 | -0,7 | potassium | 10460,0 | -0,6 |
| zinc | 97,2 | 4,6 | vanadium | 0,0 | -0,7 |

# Piper methysticum

| | |
|---|---|
| Family: | Piperaceae |
| Scientific name: | Piper methysticum |
| English name: | Awa-root |
| German name: | Kawa-Kawa |
| French name: | Poivre envirant |
| Dutch name: | Kava kava |
| Part Used: | radix |
| Extraction: | alcoholic watery |
| Titer: | 1/3 |
| Standard minimal: | |
| Comment: | |

| Sorted by element name | | | Sorted by relative difference | | |
|---|---|---|---|---|---|
| aluminium | 0,0 | -0,5 | germanium | 2,0 | 2,1 |
| arsenicum | 0,0 | -0,4 | chromium | 0,5 | 0,5 |
| cadmium | 0,0 | -0,3 | cobalt | 0,4 | 0,2 |
| calcium | 86,3 | -0,6 | cadmium | 0,0 | -0,3 |
| chromium | 0,5 | 0,5 | molybdenum | 0,0 | -0,4 |
| cobalt | 0,4 | 0,2 | selenium | 0,0 | -0,4 |
| copper | 0,6 | -0,4 | copper | 0,6 | -0,4 |
| germanium | 2,0 | 2,1 | manganese | 0,0 | -0,4 |
| iron | 1,1 | -0,5 | arsenicum | 0,0 | -0,4 |
| lead | 0,0 | -0,4 | lead | 0,0 | -0,4 |
| lithium | 0,0 | -0,4 | lithium | 0,0 | -0,4 |
| magnesium | 17,2 | -1,1 | mercury | 0,0 | -0,4 |
| manganese | 0,0 | -0,4 | aluminium | 0,0 | -0,5 |
| mercury | 0,0 | -0,4 | iron | 1,1 | -0,5 |
| molybdenum | 0,0 | -0,4 | sodium | 212,6 | -0,5 |
| nickel | 0,0 | -0,6 | calcium | 86,3 | -0,6 |
| phosphorus | 53,7 | -1,0 | nickel | 0,0 | -0,6 |
| potassium | 704,6 | -1,0 | vanadium | 0,0 | -0,7 |
| selenium | 0,0 | -0,4 | zinc | 1,5 | -0,9 |
| sodium | 212,6 | -0,5 | phosphorus | 53,7 | -1,0 |
| vanadium | 0,0 | -0,7 | potassium | 704,6 | -1,0 |
| zinc | 1,5 | -0,9 | magnesium | 17,2 | -1,1 |

# Plantago lanceolata

| | |
|---|---|
| Family: | Plantaginaceae |
| Scientific name: | Plantago lanceolata |
| English name: | Waybread |
| German name: | Wegerich |
| French name: | Plantain |
| Dutch name: | Weegbree smalbladig |
| Part Used: | folia |
| Extraction: | watery |
| Titer: | 1/4 |
| Standard minimal: | |
| Comment: | |

| Sorted by element name | | | Sorted by relative difference | | |
|---|---|---|---|---|---|
| aluminium | 13,1 | -0,3 | vanadium | 1,0 | 0,8 |
| arsenicum | 0,0 | -0,4 | cadmium | 0,2 | 0,4 |
| cadmium | 0,2 | 0,4 | calcium | 4657,0 | 0,3 |
| calcium | 4657,0 | 0,3 | sodium | 1944,0 | 0,0 |
| chromium | 0,2 | -0,2 | chromium | 0,2 | -0,2 |
| cobalt | 0,2 | -0,3 | iron | 37,3 | -0,2 |
| copper | 0,9 | -0,4 | aluminium | 13,1 | -0,3 |
| germanium | 0,1 | -0,3 | germanium | 0,1 | -0,3 |
| iron | 37,3 | -0,2 | manganese | 16,7 | -0,3 |
| lead | 0,0 | -0,4 | cobalt | 0,2 | -0,3 |
| lithium | 0,0 | -0,5 | molybdenum | 0,0 | -0,4 |
| magnesium | 955,9 | -0,6 | copper | 0,9 | -0,4 |
| manganese | 16,7 | -0,3 | selenium | 0,0 | -0,4 |
| mercury | 0,0 | -0,4 | nickel | 0,8 | -0,4 |
| molybdenum | 0,0 | -0,4 | arsenicum | 0,0 | -0,4 |
| nickel | 0,8 | -0,4 | lead | 0,0 | -0,4 |
| phosphorus | 864,5 | -0,5 | mercury | 0,0 | -0,4 |
| potassium | 13860,0 | -0,5 | lithium | 0,0 | -0,5 |
| selenium | 0,0 | -0,4 | phosphorus | 864,5 | -0,5 |
| sodium | 1944,0 | 0,0 | potassium | 13860,0 | -0,5 |
| vanadium | 1,0 | 0,8 | magnesium | 955,9 | -0,6 |
| zinc | 5,5 | -0,6 | zinc | 5,5 | -0,6 |

# Raphanus sativus niger

| | |
|---|---|
| Family: | Cruciferae |
| Scientific name: | Raphanus sativus niger |
| English name: | Black radish |
| German name: | Schwarzer Rettich |
| French name: | Radis noir, Rave |
| Dutch name: | Radijs |
| Part Used: | radix |
| Extraction: | watery |
| Titer: | 1/4 |
| Standard minimal: | |
| Comment: | |

| Sorted by element name | | | Sorted by relative difference | | |
|---|---|---|---|---|---|
| aluminium | 4,9 | -0,4 | phosphorus | 5319,0 | 2,6 |
| arsenicum | 0,0 | -0,4 | germanium | 1,6 | 1,6 |
| cadmium | 0,0 | -0,3 | potassium | 59930,0 | 1,5 |
| calcium | 2324,0 | -0,2 | vanadium | 0,8 | 0,5 |
| chromium | 0,0 | -0,6 | zinc | 16,5 | 0,0 |
| cobalt | 0,1 | -0,7 | sodium | 1888,0 | 0,0 |
| copper | 0,8 | -0,4 | calcium | 2324,0 | -0,2 |
| germanium | 1,6 | 1,6 | iron | 32,7 | -0,3 |
| iron | 32,7 | -0,3 | cadmium | 0,0 | -0,3 |
| lead | 0,0 | -0,4 | molybdenum | 0,0 | -0,4 |
| lithium | 0,0 | -0,5 | nickel | 1,1 | -0,4 |
| magnesium | 1308,0 | -0,4 | aluminium | 4,9 | -0,4 |
| manganese | 5,5 | -0,4 | selenium | 0,0 | -0,4 |
| mercury | 0,0 | -0,4 | copper | 0,8 | -0,4 |
| molybdenum | 0,0 | -0,4 | manganese | 5,5 | -0,4 |
| nickel | 1,1 | -0,4 | arsenicum | 0,0 | -0,4 |
| phosphorus | 5319,0 | 2,6 | lead | 0,0 | -0,4 |
| potassium | 59930,0 | 1,5 | magnesium | 1308,0 | -0,4 |
| selenium | 0,0 | -0,4 | mercury | 0,0 | -0,4 |
| sodium | 1888,0 | 0,0 | lithium | 0,0 | -0,5 |
| vanadium | 0,8 | 0,5 | chromium | 0,0 | -0,6 |
| zinc | 16,5 | 0,0 | cobalt | 0,1 | -0,7 |

# Rhamnus frangula

| | |
|---|---|
| Family: | Rhamnaceae |
| Scientific name: | Rhamnus frangula |
| English name: | Alder buckthorn |
| German name: | Faulbaum |
| French name: | Franguil |
| Dutch name: | Sporkehout, Vuilboom |
| Part Used: | cortex |
| Extraction: | alcoholic watery |
| Titer: | 1/3,5 |
| Standard minimal: | 1,5% glucofrangulin |
| Comment: | |

| Sorted by element name | | | Sorted by relative difference | | |
|---|---|---|---|---|---|
| aluminium | 1,0 | -0,5 | mercury | 1,6 | 2,8 |
| arsenicum | 0,7 | -0,1 | selenium | 7,2 | 1,8 |
| cadmium | 0,0 | -0,3 | manganese | 67,0 | 0,0 |
| calcium | 592,5 | -0,5 | lithium | 0,7 | 0,0 |
| chromium | 0,0 | -0,6 | nickel | 2,2 | -0,1 |
| cobalt | 0,2 | -0,3 | arsenicum | 0,7 | -0,1 |
| copper | 4,0 | -0,2 | copper | 4,0 | -0,2 |
| germanium | 0,0 | -0,5 | cadmium | 0,0 | -0,3 |
| iron | 12,1 | -0,4 | cobalt | 0,2 | -0,3 |
| lead | 0,0 | -0,4 | sodium | 825,6 | -0,4 |
| lithium | 0,7 | 0,0 | molybdenum | 0,0 | -0,4 |
| magnesium | 792,2 | -0,7 | iron | 12,1 | -0,4 |
| manganese | 67,0 | 0,0 | lead | 0,0 | -0,4 |
| mercury | 1,6 | 2,8 | aluminium | 1,0 | -0,5 |
| molybdenum | 0,0 | -0,4 | germanium | 0,0 | -0,5 |
| nickel | 2,2 | -0,1 | calcium | 592,5 | -0,5 |
| phosphorus | 349,5 | -0,8 | chromium | 0,0 | -0,6 |
| potassium | 3334,0 | -0,9 | zinc | 6,5 | -0,6 |
| selenium | 7,2 | 1,8 | magnesium | 792,2 | -0,7 |
| sodium | 825,6 | -0,4 | vanadium | 0,0 | -0,7 |
| vanadium | 0,0 | -0,7 | phosphorus | 349,5 | -0,8 |
| zinc | 6,5 | -0,6 | potassium | 3334,0 | -0,9 |

# Rhamnus purshiana

| | |
|---|---|
| Family: | Rhamnaceae |
| Scientific name: | Rhamnus purshiana |
| English name: | Buckthorn |
| German name: | Faulbaum |
| French name: | Nerprun |
| Dutch name: | Amerikaanse vuilboom |
| Part Used: | cortex |
| Extraction: | |
| Titer: | 1/4,5 |
| Standard minimal: | 4% hydroxyanthraceenheterosid |
| Comment: | |

| Sorted by element name | | | Sorted by relative difference | | |
|---|---|---|---|---|---|
| aluminium | 8,6 | -0,3 | germanium | 2,4 | 2,6 |
| arsenicum | 0,0 | -0,4 | cadmium | 0,2 | 0,4 |
| cadmium | 0,2 | 0,4 | cobalt | 0,3 | -0,1 |
| calcium | 843,7 | -0,5 | manganese | 50,0 | -0,1 |
| chromium | 0,0 | -0,6 | aluminium | 8,6 | -0,3 |
| cobalt | 0,3 | -0,1 | iron | 20,2 | -0,3 |
| copper | 1,5 | -0,4 | lithium | 0,2 | -0,3 |
| germanium | 2,4 | 2,6 | copper | 1,5 | -0,4 |
| iron | 20,2 | -0,3 | molybdenum | 0,0 | -0,4 |
| lead | 0,0 | -0,4 | selenium | 0,0 | -0,4 |
| lithium | 0,2 | -0,3 | arsenicum | 0,0 | -0,4 |
| magnesium | 956,4 | -0,6 | lead | 0,0 | -0,4 |
| manganese | 50,0 | -0,1 | mercury | 0,0 | -0,4 |
| mercury | 0,0 | -0,4 | calcium | 843,7 | -0,5 |
| molybdenum | 0,0 | -0,4 | sodium | 266,9 | -0,5 |
| nickel | 0,0 | -0,6 | chromium | 0,0 | -0,6 |
| phosphorus | 449,6 | -0,8 | magnesium | 956,4 | -0,6 |
| potassium | 7237,0 | -0,8 | nickel | 0,0 | -0,6 |
| selenium | 0,0 | -0,4 | vanadium | 0,0 | -0,7 |
| sodium | 266,9 | -0,5 | zinc | 3,9 | -0,7 |
| vanadium | 0,0 | -0,7 | phosphorus | 449,6 | -0,8 |
| zinc | 3,9 | -0,7 | potassium | 7237,0 | -0,8 |

# Rheum officinale

| | |
|---|---|
| Family: | Polygonaceae |
| Scientific name: | Rheum officinale |
| English name: | Rhubarb |
| German name: | Chinesischen Rhabarber |
| French name: | Rhubarbe |
| Dutch name: | Rabarber, chinese |
| Part Used: | radix |
| Extraction: | alcoholic watery |
| Titer: | 1/4 |
| Standard minimal: | 10% anthrachinons |
| Comment: | |

| Sorted by element name | | | Sorted by relative difference | | |
|---|---|---|---|---|---|
| aluminium | 15,8 | -0,2 | nickel | 6,1 | 0,9 |
| arsenicum | 1,7 | 0,4 | arsenicum | 1,7 | 0,4 |
| cadmium | 0,0 | -0,3 | sodium | 2581,0 | 0,2 |
| calcium | 680,0 | -0,5 | germanium | 0,3 | -0,1 |
| chromium | 0,0 | -0,6 | aluminium | 15,8 | -0,2 |
| cobalt | 0,0 | -0,9 | cadmium | 0,0 | -0,3 |
| copper | 1,4 | -0,4 | copper | 1,4 | -0,4 |
| germanium | 0,3 | -0,1 | molybdenum | 0,0 | -0,4 |
| iron | 5,7 | -0,4 | magnesium | 1448,0 | -0,4 |
| lead | 0,0 | -0,4 | selenium | 0,0 | -0,4 |
| lithium | 0,0 | -0,5 | manganese | 3,5 | -0,4 |
| magnesium | 1448,0 | -0,4 | iron | 5,7 | -0,4 |
| manganese | 3,5 | -0,4 | lead | 0,0 | -0,4 |
| mercury | 0,0 | -0,4 | mercury | 0,0 | -0,4 |
| molybdenum | 0,0 | -0,4 | lithium | 0,0 | -0,5 |
| nickel | 6,1 | 0,9 | calcium | 680,0 | -0,5 |
| phosphorus | 326,3 | -0,8 | chromium | 0,0 | -0,6 |
| potassium | 10460,0 | -0,6 | potassium | 10460,0 | -0,6 |
| selenium | 0,0 | -0,4 | vanadium | 0,0 | -0,7 |
| sodium | 2581,0 | 0,2 | zinc | 3,0 | -0,8 |
| vanadium | 0,0 | -0,7 | phosphorus | 326,3 | -0,8 |
| zinc | 3,0 | -0,8 | cobalt | 0,0 | -0,9 |

# Ribes nigrum folia

| | |
|---|---|
| Family: | Saxifragaceae |
| Scientific name: | Ribes nigrum folia |
| English name: | Blackbarry |
| German name: | |
| French name: | |
| Dutch name: | Zwarte bes |
| Part Used: | folia |
| Extraction: | alcoholic watery |
| Titer: | 1/5 |
| Standard minimal: | |
| Comment: | |

| Sorted by element name | | | Sorted by relative difference | | |
|---|---|---|---|---|---|
| aluminium | 13,7 | -0,3 | mercury | 0,6 | 0,8 |
| arsenicum | 0,0 | -0,4 | vanadium | 0,9 | 0,6 |
| cadmium | 0,0 | -0,3 | magnesium | 3152,0 | 0,5 |
| calcium | 3651,0 | 0,1 | lithium | 1,3 | 0,3 |
| chromium | 0,0 | -0,6 | calcium | 3651,0 | 0,1 |
| cobalt | 0,1 | -0,5 | nickel | 2,4 | 0,0 |
| copper | 3,3 | -0,2 | copper | 3,3 | -0,2 |
| germanium | 0,0 | -0,5 | aluminium | 13,7 | -0,3 |
| iron | 30,0 | -0,3 | cadmium | 0,0 | -0,3 |
| lead | 0,0 | -0,4 | iron | 30,0 | -0,3 |
| lithium | 1,3 | 0,3 | potassium | 18250,0 | -0,3 |
| magnesium | 3152,0 | 0,5 | manganese | 19,0 | -0,3 |
| manganese | 19,0 | -0,3 | molybdenum | 0,0 | -0,4 |
| mercury | 0,6 | 0,8 | phosphorus | 998,1 | -0,4 |
| molybdenum | 0,0 | -0,4 | selenium | 0,0 | -0,4 |
| nickel | 2,4 | 0,0 | arsenicum | 0,0 | -0,4 |
| phosphorus | 998,1 | -0,4 | lead | 0,0 | -0,4 |
| potassium | 18250,0 | -0,3 | germanium | 0,0 | -0,5 |
| selenium | 0,0 | -0,4 | cobalt | 0,1 | -0,5 |
| sodium | 80,7 | -0,6 | zinc | 6,7 | -0,6 |
| vanadium | 0,9 | 0,6 | chromium | 0,0 | -0,6 |
| zinc | 6,7 | -0,6 | sodium | 80,7 | -0,6 |

# Ribes nigrum fructus

| | |
|---|---|
| Family: | Saxifragaceae |
| Scientific name: | Ribes nigrum fructus |
| English name: | Blackbarry |
| German name: | |
| French name: | |
| Dutch name: | Zwarte bes |
| Part Used: | fructus |
| Extraction: | watery |
| Titer: | 1/6,5 |
| Standard minimal: | |
| Comment: | |

| Sorted by element name | | | Sorted by relative difference | | |
|---|---|---|---|---|---|
| aluminium | 10,2 | -0,3 | chromium | 0,3 | 0,1 |
| arsenicum | 0,0 | -0,4 | lithium | 0,8 | 0,0 |
| cadmium | 0,0 | -0,3 | cadmium | 0,0 | -0,3 |
| calcium | 988,1 | -0,5 | iron | 28,5 | -0,3 |
| chromium | 0,3 | 0,1 | aluminium | 10,2 | -0,3 |
| cobalt | 0,0 | -0,9 | molybdenum | 0,0 | -0,4 |
| copper | 1,0 | -0,4 | copper | 1,0 | -0,4 |
| germanium | 0,0 | -0,5 | manganese | 6,6 | -0,4 |
| iron | 28,5 | -0,3 | selenium | 0,0 | -0,4 |
| lead | 0,0 | -0,4 | arsenicum | 0,0 | -0,4 |
| lithium | 0,8 | 0,0 | lead | 0,0 | -0,4 |
| magnesium | 344,1 | -0,9 | mercury | 0,0 | -0,4 |
| manganese | 6,6 | -0,4 | phosphorus | 906,9 | -0,4 |
| mercury | 0,0 | -0,4 | calcium | 988,1 | -0,5 |
| molybdenum | 0,0 | -0,4 | germanium | 0,0 | -0,5 |
| nickel | 0,0 | -0,6 | sodium | 242,9 | -0,5 |
| phosphorus | 906,9 | -0,4 | nickel | 0,0 | -0,6 |
| potassium | 9168,0 | -0,7 | zinc | 5,0 | -0,7 |
| selenium | 0,0 | -0,4 | potassium | 9168,0 | -0,7 |
| sodium | 242,9 | -0,5 | vanadium | 0,0 | -0,7 |
| vanadium | 0,0 | -0,7 | cobalt | 0,0 | -0,9 |
| zinc | 5,0 | -0,7 | magnesium | 344,1 | -0,9 |

# Rosa canina

| | |
|---|---|
| Family: | Rosaceae |
| Scientific name: | Rose canina |
| English name: | Dog rose |
| German name: | |
| French name: | |
| Dutch name: | Hondsroos |
| Part Used: | fructus |
| Extraction: | watery |
| Titer: | 1/3 |
| Standard minimal: | |
| Comment: | |

| Sorted by element name | | | Sorted by relative difference | | |
|---|---|---|---|---|---|
| aluminium | 1,8 | -0,4 | lead | 2,8 | 2,4 |
| arsenicum | 0,3 | -0,3 | germanium | 1,7 | 1,7 |
| cadmium | 0,3 | 0,7 | cadmium | 0,3 | 0,7 |
| calcium | 746,0 | -0,5 | molybdenum | 0,2 | -0,1 |
| chromium | 0,0 | -0,6 | lithium | 0,4 | -0,2 |
| cobalt | 0,1 | -0,6 | arsenicum | 0,3 | -0,3 |
| copper | 0,2 | -0,5 | manganese | 14,7 | -0,3 |
| germanium | 1,7 | 1,7 | selenium | 0,0 | -0,4 |
| iron | 4,2 | -0,4 | sodium | 580,0 | -0,4 |
| lead | 2,8 | 2,4 | aluminium | 1,8 | -0,4 |
| lithium | 0,4 | -0,2 | iron | 4,2 | -0,4 |
| magnesium | 736,5 | -0,7 | mercury | 0,0 | -0,4 |
| manganese | 14,7 | -0,3 | copper | 0,2 | -0,5 |
| mercury | 0,0 | -0,4 | calcium | 746,0 | -0,5 |
| molybdenum | 0,2 | -0,1 | chromium | 0,0 | -0,6 |
| nickel | 0,0 | -0,6 | cobalt | 0,1 | -0,6 |
| phosphorus | 358,4 | -0,8 | nickel | 0,0 | -0,6 |
| potassium | 7330,0 | -0,8 | vanadium | 0,0 | -0,7 |
| selenium | 0,0 | -0,4 | magnesium | 736,5 | -0,7 |
| sodium | 580,0 | -0,4 | potassium | 7330,0 | -0,8 |
| vanadium | 0,0 | -0,7 | zinc | 2,7 | -0,8 |
| zinc | 2,7 | -0,8 | phosphorus | 358,4 | -0,8 |

# Rosmarinus officinalis

Family:              Labiatae
Scientific name:     Rosmarinus officinalis
English name:        Rosemary
German name:         Rosmarin
French name:         Rosmarin, Somarin
Dutch name:          Rozemarijn
Part Used:           folia
Extraction:          watery
Titer:               1/5
Standard minimal:    3% rosmarin acid, 2% etheric oil
Comment:

| Sorted by element name | | | Sorted by relative difference | | |
|---|---|---|---|---|---|
| aluminium | 15,8 | -0,2 | molybdenum | 2,8 | 3,7 |
| arsenicum | 0,2 | -0,3 | nickel | 11,0 | 2,1 |
| cadmium | 0,0 | -0,3 | calcium | 11690,0 | 1,7 |
| calcium | 11690,0 | 1,7 | selenium | 5,1 | 1,2 |
| chromium | 0,1 | -0,4 | mercury | 0,6 | 0,8 |
| cobalt | 0,0 | -0,9 | vanadium | 0,8 | 0,5 |
| copper | 1,6 | -0,3 | magnesium | 3004,0 | 0,4 |
| germanium | 0,0 | -0,5 | potassium | 34290,0 | 0,4 |
| iron | 61,7 | -0,1 | lithium | 1,3 | 0,3 |
| lead | 0,1 | -0,3 | sodium | 1861,0 | 0,0 |
| lithium | 1,3 | 0,3 | iron | 61,7 | -0,1 |
| magnesium | 3004,0 | 0,4 | aluminium | 15,8 | -0,2 |
| manganese | 21,9 | -0,3 | cadmium | 0,0 | -0,3 |
| mercury | 0,6 | 0,8 | manganese | 21,9 | -0,3 |
| molybdenum | 2,8 | 3,7 | phosphorus | 1072,0 | -0,3 |
| nickel | 11,0 | 2,1 | arsenicum | 0,2 | -0,3 |
| phosphorus | 1072,0 | -0,3 | copper | 1,6 | -0,3 |
| potassium | 34290,0 | 0,4 | lead | 0,1 | -0,3 |
| selenium | 5,1 | 1,2 | chromium | 0,1 | -0,4 |
| sodium | 1861,0 | 0,0 | germanium | 0,0 | -0,5 |
| vanadium | 0,8 | 0,5 | zinc | 7,0 | -0,5 |
| zinc | 7,0 | -0,5 | cobalt | 0,0 | -0,9 |

# Ruscus aculeatus

| | |
|---|---|
| Family: | Liliaceae |
| Scientific name: | Ruscus aculeatus |
| English name: | |
| German name: | |
| French name: | Epine de rat |
| Dutch name: | Muisdoorn |
| Part Used: | rhizoma |
| Extraction: | alcoholic watery |
| Titer: | 1/4 |
| Standard minimal: | |
| Comment: | |

| Sorted by element name | | | Sorted by relative difference | | |
|---|---|---|---|---|---|
| aluminium | 2,3 | -0,4 | germanium | 3,3 | 3,8 |
| arsenicum | 0,0 | -0,4 | cobalt | 1,1 | 2,0 |
| cadmium | 0,0 | -0,1 | chromium | 0,3 | 0,1 |
| calcium | 414,8 | -0,6 | cadmium | 0,0 | -0,1 |
| chromium | 0,3 | 0,1 | copper | 2,4 | -0,3 |
| cobalt | 1,1 | 2,0 | manganese | 10,5 | -0,4 |
| copper | 2,4 | -0,3 | molybdenum | 0,0 | -0,4 |
| germanium | 3,3 | 3,8 | sodium | 732,1 | -0,4 |
| iron | 9,3 | -0,4 | selenium | 0,0 | -0,4 |
| lead | 0,0 | -0,4 | iron | 9,3 | -0,4 |
| lithium | 0,0 | -0,5 | nickel | 0,9 | -0,4 |
| magnesium | 228,2 | -1,0 | aluminium | 2,3 | -0,4 |
| manganese | 10,5 | -0,4 | arsenicum | 0,0 | -0,4 |
| mercury | 0,0 | -0,4 | lead | 0,0 | -0,4 |
| molybdenum | 0,0 | -0,4 | mercury | 0,0 | -0,4 |
| nickel | 0,9 | -0,4 | lithium | 0,0 | -0,5 |
| phosphorus | 327,2 | -0,8 | calcium | 414,8 | -0,6 |
| potassium | 8145,0 | -0,7 | potassium | 8145,0 | -0,7 |
| selenium | 0,0 | -0,4 | vanadium | 0,0 | -0,7 |
| sodium | 732,1 | -0,4 | zinc | 3,3 | -0,8 |
| vanadium | 0,0 | -0,7 | phosphorus | 327,2 | -0,8 |
| zinc | 3,3 | -0,8 | magnesium | 228,2 | -1,0 |

# Salix alba

| | |
|---|---|
| Family: | Salicaceae |
| Scientific name: | Salix alba |
| English name: | White willow |
| German name: | Weisse weide |
| French name: | Saule blanc |
| Dutch name: | Wilg witte |
| Part Used: | cortex |
| Extraction: | watery |
| Titer: | 1/7 |
| Standard minimal: | 7% salicin |
| Comment: | |

| Sorted by element name | | | Sorted by relative difference | | |
|---|---|---|---|---|---|
| aluminium | 32,4 | 0,0 | vanadium | 2,6 | 3,1 |
| arsenicum | 0,0 | -0,4 | cobalt | 1,3 | 2,5 |
| cadmium | 0,0 | -0,3 | zinc | 59,8 | 2,5 |
| calcium | 8850,0 | 1,1 | calcium | 8850,0 | 1,1 |
| chromium | 0,4 | 0,2 | manganese | 170,0 | 0,7 |
| cobalt | 1,3 | 2,5 | iron | 134,0 | 0,4 |
| copper | 2,7 | -0,3 | potassium | 34050,0 | 0,4 |
| germanium | 0,0 | -0,5 | magnesium | 2824,0 | 0,3 |
| iron | 134,0 | 0,4 | phosphorus | 1916,0 | 0,3 |
| lead | 0,0 | -0,4 | chromium | 0,4 | 0,2 |
| lithium | 0,0 | -0,5 | nickel | 3,0 | 0,1 |
| magnesium | 2824,0 | 0,3 | aluminium | 32,4 | 0,0 |
| manganese | 170,0 | 0,7 | cadmium | 0,0 | -0,3 |
| mercury | 0,0 | -0,4 | copper | 2,7 | -0,3 |
| molybdenum | 0,0 | -0,4 | molybdenum | 0,0 | -0,4 |
| nickel | 3,0 | 0,1 | sodium | 791,8 | -0,4 |
| phosphorus | 1916,0 | 0,3 | selenium | 0,0 | -0,4 |
| potassium | 34050,0 | 0,4 | arsenicum | 0,0 | -0,4 |
| selenium | 0,0 | -0,4 | lead | 0,0 | -0,4 |
| sodium | 791,8 | -0,4 | mercury | 0,0 | -0,4 |
| vanadium | 2,6 | 3,1 | lithium | 0,0 | -0,5 |
| zinc | 59,8 | 2,5 | germanium | 0,0 | -0,5 |

# Salvia officinalis

| | |
|---|---|
| Family: | Labiatae |
| Scientific name: | Salvia officinalis |
| English name: | Garden sage |
| German name: | Gartensalbei |
| French name: | Suage officinale |
| Dutch name: | Salie |
| Part Used: | folia |
| Extraction: | watery |
| Titer: | 1/5,5 |
| Standard minimal: | 3% rosmarin acid, 3% etheric oil |
| Comment: | |

| Sorted by element name | | | Sorted by relative difference | | |
|---|---|---|---|---|---|
| aluminium | 18,7 | -0,2 | calcium | 7311,0 | 0,8 |
| arsenicum | 0,0 | -0,4 | vanadium | 1,0 | 0,8 |
| cadmium | 0,1 | -0,1 | magnesium | 3153,0 | 0,5 |
| calcium | 7311,0 | 0,8 | germanium | 0,7 | 0,4 |
| chromium | 0,1 | -0,4 | lithium | 0,7 | 0,0 |
| cobalt | 0,3 | -0,1 | cadmium | 0,1 | -0,1 |
| copper | 0,4 | -0,4 | cobalt | 0,3 | -0,1 |
| germanium | 0,7 | 0,4 | aluminium | 18,7 | -0,2 |
| iron | 29,2 | -0,3 | molybdenum | 0,1 | -0,2 |
| lead | 0,0 | -0,4 | manganese | 28,6 | -0,2 |
| lithium | 0,7 | 0,0 | iron | 29,2 | -0,3 |
| magnesium | 3153,0 | 0,5 | potassium | 17940,0 | -0,3 |
| manganese | 28,6 | -0,2 | chromium | 0,1 | -0,4 |
| mercury | 0,0 | -0,4 | selenium | 0,0 | -0,4 |
| molybdenum | 0,1 | -0,2 | sodium | 625,5 | -0,4 |
| nickel | 0,0 | -0,6 | copper | 0,4 | -0,4 |
| phosphorus | 801,1 | -0,5 | arsenicum | 0,0 | -0,4 |
| potassium | 17940,0 | -0,3 | lead | 0,0 | -0,4 |
| selenium | 0,0 | -0,4 | mercury | 0,0 | -0,4 |
| sodium | 625,5 | -0,4 | phosphorus | 801,1 | -0,5 |
| vanadium | 1,0 | 0,8 | zinc | 6,6 | -0,6 |
| zinc | 6,6 | -0,6 | nickel | 0,0 | -0,6 |

# Sarsaparilla

| | |
|---|---|
| Family: | Liliaceae |
| Scientific name: | Sarsaparilla |
| English name: | Sarsaparilla |
| German name: | Sarsaparilla |
| French name: | Sarsaparilla |
| Dutch name: | Sarsaparilla |
| Part Used: | radix |
| Extraction: | alcoholic watery |
| Titer: | 1/4 |
| Standard minimal: | |
| Comment: | |

| Sorted by element name | | | Sorted by relative difference | | |
|---|---|---|---|---|---|
| aluminium | 7,3 | -0,4 | molybdenum | 3,3 | 4,4 |
| arsenicum | 0,1 | -0,4 | cobalt | 0,4 | 0,2 |
| cadmium | 0,0 | -0,3 | vanadium | 0,6 | 0,2 |
| calcium | 1996,0 | -0,3 | potassium | 27100,0 | 0,1 |
| chromium | 0,0 | -0,6 | calcium | 1996,0 | -0,3 |
| cobalt | 0,4 | 0,2 | cadmium | 0,0 | -0,3 |
| copper | 0,7 | -0,4 | sodium | 1106,0 | -0,3 |
| germanium | 0,0 | -0,5 | iron | 26,7 | -0,3 |
| iron | 26,7 | -0,3 | manganese | 18,7 | -0,3 |
| lead | 0,0 | -0,4 | aluminium | 7,3 | -0,4 |
| lithium | 0,0 | -0,5 | zinc | 9,9 | -0,4 |
| magnesium | 1225,0 | -0,5 | arsenicum | 0,1 | -0,4 |
| manganese | 18,7 | -0,3 | phosphorus | 988,2 | -0,4 |
| mercury | 0,0 | -0,4 | selenium | 0,0 | -0,4 |
| molybdenum | 3,3 | 4,4 | copper | 0,7 | -0,4 |
| nickel | 0,3 | -0,6 | lead | 0,0 | -0,4 |
| phosphorus | 988,2 | -0,4 | mercury | 0,0 | -0,4 |
| potassium | 27100,0 | 0,1 | lithium | 0,0 | -0,5 |
| selenium | 0,0 | -0,4 | germanium | 0,0 | -0,5 |
| sodium | 1106,0 | -0,3 | magnesium | 1225,0 | -0,5 |
| vanadium | 0,6 | 0,2 | nickel | 0,3 | -0,6 |
| zinc | 9,9 | -0,4 | chromium | 0,0 | -0,6 |

# Senna

| | |
|---|---|
| Family: | Leguminosae |
| Scientific name: | Cassia angustifolia |
| English name: | Senna |
| German name: | Sennesblätterstrauch |
| French name: | Séné |
| Dutch name: | Senna |
| Part Used: | fructus |
| Extraction: | watery |
| Titer: | 1/4 |
| Standard minimal: | 7% sennosids |
| Comment: | |

| Sorted by element name | | | Sorted by relative difference | | |
|---|---|---|---|---|---|
| aluminium | 4,9 | -0,4 | zinc | 41,6 | 1,4 |
| arsenicum | 0,0 | -0,4 | cadmium | 0,3 | 0,6 |
| cadmium | 0,3 | 0,6 | nickel | 3,6 | 0,3 |
| calcium | 332,2 | -0,6 | copper | 5,4 | -0,1 |
| chromium | 0,2 | -0,2 | potassium | 21990,0 | -0,1 |
| cobalt | 0,2 | -0,4 | sodium | 1544,0 | -0,1 |
| copper | 5,4 | -0,1 | chromium | 0,2 | -0,2 |
| germanium | 0,0 | -0,5 | magnesium | 1523,0 | -0,3 |
| iron | 9,8 | -0,4 | cobalt | 0,2 | -0,4 |
| lead | 0,0 | -0,4 | molybdenum | 0,0 | -0,4 |
| lithium | 0,0 | -0,5 | aluminium | 4,9 | -0,4 |
| magnesium | 1523,0 | -0,3 | selenium | 0,0 | -0,4 |
| manganese | 3,9 | -0,4 | iron | 9,8 | -0,4 |
| mercury | 0,0 | -0,4 | manganese | 3,9 | -0,4 |
| molybdenum | 0,0 | -0,4 | arsenicum | 0,0 | -0,4 |
| nickel | 3,6 | 0,3 | lead | 0,0 | -0,4 |
| phosphorus | 771,7 | -0,5 | mercury | 0,0 | -0,4 |
| potassium | 21990,0 | -0,1 | lithium | 0,0 | -0,5 |
| selenium | 0,0 | -0,4 | germanium | 0,0 | -0,5 |
| sodium | 1544,0 | -0,1 | phosphorus | 771,7 | -0,5 |
| vanadium | 0,0 | -0,7 | calcium | 332,2 | -0,6 |
| zinc | 41,6 | 1,4 | vanadium | 0,0 | -0,7 |

# Solidago virga aurea

| | |
|---|---|
| Family: | Compositae |
| Scientific name: | Solidago virga aurea |
| English name: | Golden rod |
| German name: | Goldrute |
| French name: | Verge d'or |
| Dutch name: | Gulden roede |
| Part Used: | herba |
| Extraction: | alcoholic watery |
| Titer: | 1/6 |
| Standard minimal: | 6% flavonoids |
| Comment: | |

| Sorted by element name | | | Sorted by relative difference | | |
|---|---|---|---|---|---|
| aluminium | 3,1 | -0,4 | lead | 2,5 | 2,1 |
| arsenicum | 0,4 | -0,2 | potassium | 44890,0 | 0,8 |
| cadmium | 0,0 | -0,3 | mercury | 0,6 | 0,8 |
| calcium | 2731,0 | -0,1 | vanadium | 1,0 | 0,8 |
| chromium | 0,0 | -0,6 | phosphorus | 2519,0 | 0,7 |
| cobalt | 0,1 | -0,6 | magnesium | 3354,0 | 0,6 |
| copper | 12,7 | 0,5 | zinc | 25,7 | 0,5 |
| germanium | 0,0 | -0,5 | copper | 12,7 | 0,5 |
| iron | 12,9 | -0,4 | calcium | 2731,0 | -0,1 |
| lead | 2,5 | 2,1 | manganese | 37,3 | -0,2 |
| lithium | 0,3 | -0,3 | arsenicum | 0,4 | -0,2 |
| magnesium | 3354,0 | 0,6 | cadmium | 0,0 | -0,3 |
| manganese | 37,3 | -0,2 | nickel | 1,5 | -0,3 |
| mercury | 0,6 | 0,8 | lithium | 0,3 | -0,3 |
| molybdenum | 0,0 | -0,4 | molybdenum | 0,0 | -0,4 |
| nickel | 1,5 | -0,3 | iron | 12,9 | -0,4 |
| phosphorus | 2519,0 | 0,7 | selenium | 0,0 | -0,4 |
| potassium | 44890,0 | 0,8 | aluminium | 3,1 | -0,4 |
| selenium | 0,0 | -0,4 | germanium | 0,0 | -0,5 |
| sodium | 335,9 | -0,5 | sodium | 335,9 | -0,5 |
| vanadium | 1,0 | 0,8 | chromium | 0,0 | -0,6 |
| zinc | 25,7 | 0,5 | cobalt | 0,1 | -0,6 |

# Spiraea ulmaria

| | |
|---|---|
| Family: | Rosaceae |
| Scientific name: | Spiraea ulmaria |
| English name: | Meadowsweet |
| German name: | |
| French name: | Reine des pres |
| Dutch name: | Moerasspiraea |
| Part Used: | herba |
| Extraction: | alcoholic watery |
| Titer: | 1/5 |
| Standard minimal: | 1% salicin and derivates |
| Comment: | |

| Sorted by element name | | | Sorted by relative difference | | |
|---|---|---|---|---|---|
| aluminium | 0,2 | -0,5 | lithium | 9,7 | 5,5 |
| arsenicum | 0,0 | -0,4 | selenium | 5,6 | 1,3 |
| cadmium | 0,1 | 0,1 | magnesium | 3209,0 | 0,5 |
| calcium | 1251,0 | -0,4 | molybdenum | 0,6 | 0,5 |
| chromium | 0,0 | -0,6 | cadmium | 0,1 | 0,1 |
| cobalt | 0,0 | -0,9 | copper | 6,7 | 0,0 |
| copper | 6,7 | 0,0 | nickel | 2,5 | 0,0 |
| germanium | 0,0 | -0,5 | phosphorus | 1531,0 | 0,0 |
| iron | 6,9 | -0,4 | zinc | 14,9 | -0,1 |
| lead | 0,0 | -0,4 | potassium | 21660,0 | -0,2 |
| lithium | 9,7 | 5,5 | mercury | 0,1 | -0,2 |
| magnesium | 3209,0 | 0,5 | manganese | 20,4 | -0,3 |
| manganese | 20,4 | -0,3 | sodium | 908,4 | -0,3 |
| mercury | 0,1 | -0,2 | calcium | 1251,0 | -0,4 |
| molybdenum | 0,6 | 0,5 | iron | 6,9 | -0,4 |
| nickel | 2,5 | 0,0 | arsenicum | 0,0 | -0,4 |
| phosphorus | 1531,0 | 0,0 | lead | 0,0 | -0,4 |
| potassium | 21660,0 | -0,2 | aluminium | 0,2 | -0,5 |
| selenium | 5,6 | 1,3 | germanium | 0,0 | -0,5 |
| sodium | 908,4 | -0,3 | chromium | 0,0 | -0,6 |
| vanadium | 0,0 | -0,7 | vanadium | 0,0 | -0,7 |
| zinc | 14,9 | -0,1 | cobalt | 0,0 | -0,9 |

# Spirulina neb          I

Family:                 Oscillatoriaceae
Scientific name:        Spirulina platensis
English name:           Spirulina
German name:            Spirulina
French name:            Spirulina
Dutch name:             Spirulina
Part Used:
Extraction:
Titer:                  1/
Standard minimal:
Comment:

| Sorted by element name | | | Sorted by relative difference | | |
|---|---|---|---|---|---|
| aluminium | 6,6 | -0,4 | molybdenum | 0,7 | 0,7 |
| arsenicum | 0,0 | -0,4 | sodium | 3305,0 | 0,4 |
| cadmium | 0,0 | -0,3 | vanadium | 0,6 | 0,2 |
| calcium | 1758,0 | -0,3 | potassium | 27212,0 | 0,1 |
| chromium | 0,2 | -0,2 | chromium | 0,2 | -0,2 |
| cobalt | 0,1 | -0,7 | lithium | 0,5 | -0,2 |
| copper | 2,2 | -0,3 | cadmium | 0,0 | -0,3 |
| germanium | 0,0 | -0,5 | iron | 30,2 | -0,3 |
| iron | 30,2 | -0,3 | phosphorus | 1154,0 | -0,3 |
| lead | 0,0 | -0,4 | calcium | 1758,0 | -0,3 |
| lithium | 0,5 | -0,2 | copper | 2,2 | -0,3 |
| magnesium | 823,7 | -0,7 | aluminium | 6,6 | -0,4 |
| manganese | 5,4 | -0,4 | selenium | 0,0 | -0,4 |
| mercury | 0,0 | -0,4 | manganese | 5,4 | -0,4 |
| molybdenum | 0,7 | 0,7 | arsenicum | 0,0 | -0,4 |
| nickel | 0,0 | -0,6 | lead | 0,0 | -0,4 |
| phosphorus | 1154,0 | -0,3 | mercury | 0,0 | -0,4 |
| potassium | 27212,0 | 0,1 | germanium | 0,0 | -0,5 |
| selenium | 0,0 | -0,4 | zinc | 5,8 | -0,6 |
| sodium | 3305,0 | 0,4 | nickel | 0,0 | -0,6 |
| vanadium | 0,6 | 0,2 | magnesium | 823,7 | -0,7 |
| zinc | 5,8 | -0,6 | cobalt | 0,1 | -0,7 |

# Spirulina poud

| | |
|---|---|
| Family: | Oscillatoriaceae |
| Scientific name: | Spirulinaplatensis |
| English name: | Spirulina |
| German name: | Spirulina |
| French name: | Spirulina |
| Dutch name: | Spirulina |
| Part Used: | |
| Extraction: | |
| Titer: | 1/ |
| Standard minimal: | |
| Comment: | |

| Sorted by element name | | | Sorted by relative difference | | |
|---|---|---|---|---|---|
| aluminium | 9,0 | -0,3 | sodium | 12390,0 | 3,1 |
| arsenicum | 6,7 | 3,0 | phosphorus | 5976,0 | 3,0 |
| cadmium | 0,3 | 0,6 | arsenicum | 6,7 | 3,0 |
| calcium | 1607,0 | -0,3 | nickel | 11,2 | 2,2 |
| chromium | 0,2 | -0,3 | iron | 315,5 | 1,7 |
| cobalt | 0,0 | -0,9 | germanium | 1,1 | 1,0 |
| copper | 0,9 | -0,4 | cadmium | 0,3 | 0,6 |
| germanium | 1,1 | 1,0 | magnesium | 2506,0 | 0,2 |
| iron | 315,5 | 1,7 | zinc | 16,6 | 0,0 |
| lead | 0,0 | -0,4 | manganese | 35,3 | -0,2 |
| lithium | 0,4 | -0,2 | lithium | 0,4 | -0,2 |
| magnesium | 2506,0 | 0,2 | chromium | 0,2 | -0,3 |
| manganese | 35,3 | -0,2 | aluminium | 9,0 | -0,3 |
| mercury | 0,0 | -0,4 | calcium | 1607,0 | -0,3 |
| molybdenum | 0,0 | -0,4 | molybdenum | 0,0 | -0,4 |
| nickel | 11,2 | 2,2 | copper | 0,9 | -0,4 |
| phosphorus | 5976,0 | 3,0 | selenium | 0,0 | -0,4 |
| potassium | 12780,0 | -0,5 | lead | 0,0 | -0,4 |
| selenium | 0,0 | -0,4 | mercury | 0,0 | -0,4 |
| sodium | 12390,0 | 3,1 | potassium | 12780,0 | -0,5 |
| vanadium | 0,0 | -0,7 | vanadium | 0,0 | -0,7 |
| zinc | 16,6 | 0,0 | cobalt | 0,0 | -0,9 |

# Taraxacum

| | |
|---|---|
| Family: | Compositae |
| Scientific name: | Taraxacum |
| English name: | Dandelion |
| German name: | Löwenzahn |
| French name: | Pinsenlit |
| Dutch name: | Paardebloem |
| Part Used: | herba radix |
| Extraction: | watery |
| Titer: | 1/2,5 |
| Standard minimal: | |
| Comment: | |

| Sorted by element name | | | | Sorted by relative difference | | |
|---|---|---|---|---|---|---|
| aluminium | 11,5 | -0,3 | | lead | 3,3 | 2,9 |
| arsenicum | 0,0 | -0,4 | | sodium | 8746,0 | 2,0 |
| cadmium | 0,1 | 0,0 | | germanium | 1,7 | 1,7 |
| calcium | 3041,0 | 0,0 | | cobalt | 1,0 | 1,7 |
| chromium | 0,2 | -0,2 | | vanadium | 1,3 | 1,2 |
| cobalt | 1,0 | 1,7 | | mercury | 0,6 | 0,8 |
| copper | 1,2 | -0,4 | | magnesium | 3581,0 | 0,7 |
| germanium | 1,7 | 1,7 | | lithium | 1,5 | 0,5 |
| iron | 45,6 | -0,2 | | cadmium | 0,1 | 0,0 |
| lead | 3,3 | 2,9 | | calcium | 3041,0 | 0,0 |
| lithium | 1,5 | 0,5 | | chromium | 0,2 | -0,2 |
| magnesium | 3581,0 | 0,7 | | iron | 45,6 | -0,2 |
| manganese | 19,8 | -0,3 | | aluminium | 11,5 | -0,3 |
| mercury | 0,6 | 0,8 | | manganese | 19,8 | -0,3 |
| molybdenum | 0,0 | -0,4 | | nickel | 1,3 | -0,3 |
| nickel | 1,3 | -0,3 | | molybdenum | 0,0 | -0,4 |
| phosphorus | 630,8 | -0,6 | | copper | 1,2 | -0,4 |
| potassium | 14100,0 | -0,5 | | selenium | 0,0 | -0,4 |
| selenium | 0,0 | -0,4 | | arsenicum | 0,0 | -0,4 |
| sodium | 8746,0 | 2,0 | | potassium | 14100,0 | -0,5 |
| vanadium | 1,3 | 1,2 | | phosphorus | 630,8 | -0,6 |
| zinc | 4,4 | -0,7 | | zinc | 4,4 | -0,7 |

# Thymus vulgaris

| | |
|---|---|
| Family: | Labiatae |
| Scientific name: | Thymus vulgaris |
| English name: | Thyme |
| German name: | Quendel, Feldthymian |
| French name: | Thym |
| Dutch name: | Thijm |
| Part Used: | herba |
| Extraction: | watery |
| Titer: | 1/4 |
| Standard minimal: | 1% etheric oil |
| Comment: | |

| Sorted by element name | | | Sorted by relative difference | | |
|---|---|---|---|---|---|
| aluminium | 20,3 | -0,2 | vanadium | 2,4 | 2,8 |
| arsenicum | 0,0 | -0,4 | selenium | 6,4 | 1,6 |
| cadmium | 0,1 | 0,0 | calcium | 9903,0 | 1,3 |
| calcium | 9903,0 | 1,3 | magnesium | 3152,0 | 0,5 |
| chromium | 0,4 | 0,3 | chromium | 0,4 | 0,3 |
| cobalt | 0,1 | -0,6 | sodium | 2054,0 | 0,0 |
| copper | 2,5 | -0,3 | cadmium | 0,1 | 0,0 |
| germanium | 0,0 | -0,5 | potassium | 24550,0 | 0,0 |
| iron | 45,8 | -0,2 | manganese | 59,1 | 0,0 |
| lead | 0,0 | -0,4 | phosphorus | 1358,0 | -0,1 |
| lithium | 0,2 | -0,3 | zinc | 14,2 | -0,1 |
| magnesium | 3152,0 | 0,5 | aluminium | 20,3 | -0,2 |
| manganese | 59,1 | 0,0 | iron | 45,8 | -0,2 |
| mercury | 0,0 | -0,4 | copper | 2,5 | -0,3 |
| molybdenum | 0,0 | -0,4 | lithium | 0,2 | -0,3 |
| nickel | 0,5 | -0,5 | molybdenum | 0,0 | -0,4 |
| phosphorus | 1358,0 | -0,1 | arsenicum | 0,0 | -0,4 |
| potassium | 24550,0 | 0,0 | lead | 0,0 | -0,4 |
| selenium | 6,4 | 1,6 | mercury | 0,0 | -0,4 |
| sodium | 2054,0 | 0,0 | germanium | 0,0 | -0,5 |
| vanadium | 2,4 | 2,8 | nickel | 0,5 | -0,5 |
| zinc | 14,2 | -0,1 | cobalt | 0,1 | -0,6 |

# Tilia alburnum

| | |
|---|---|
| Family: | Tiliaceae |
| Scientific name: | Tilia alburnum |
| English name: | Lime tree |
| German name: | Lindenbaum |
| French name: | Tilleul |
| Dutch name: | Linde |
| Part Used: | cortex |
| Extraction: | watery |
| Titer: | 1/10 |
| Standard minimal: | |
| Comment: | |

| Sorted by element name | | | Sorted by relative difference | | |
|---|---|---|---|---|---|
| aluminium | 31,5 | 0,0 | vanadium | 1,0 | 0,8 |
| arsenicum | 0,0 | -0,4 | aluminium | 31,5 | 0,0 |
| cadmium | 0,0 | -0,3 | calcium | 3306,0 | 0,0 |
| calcium | 3306,0 | 0,0 | chromium | 0,2 | -0,2 |
| chromium | 0,2 | -0,2 | nickel | 1,9 | -0,2 |
| cobalt | 0,0 | -0,9 | copper | 3,2 | -0,2 |
| copper | 3,2 | -0,2 | iron | 35,9 | -0,2 |
| germanium | 0,0 | -0,5 | cadmium | 0,0 | -0,3 |
| iron | 35,9 | -0,2 | manganese | 22,5 | -0,3 |
| lead | 0,0 | -0,4 | lithium | 0,2 | -0,3 |
| lithium | 0,2 | -0,3 | molybdenum | 0,0 | -0,4 |
| magnesium | 850,4 | -0,7 | selenium | 0,0 | -0,4 |
| manganese | 22,5 | -0,3 | sodium | 631,5 | -0,4 |
| mercury | 0,0 | -0,4 | arsenicum | 0,0 | -0,4 |
| molybdenum | 0,0 | -0,4 | lead | 0,0 | -0,4 |
| nickel | 1,9 | -0,2 | mercury | 0,0 | -0,4 |
| phosphorus | 740,9 | -0,6 | germanium | 0,0 | -0,5 |
| potassium | 10400,0 | -0,6 | phosphorus | 740,9 | -0,6 |
| selenium | 0,0 | -0,4 | potassium | 10400,0 | -0,6 |
| sodium | 631,5 | -0,4 | zinc | 5,1 | -0,7 |
| vanadium | 1,0 | 0,8 | magnesium | 850,4 | -0,7 |
| zinc | 5,1 | -0,7 | cobalt | 0,0 | -0,9 |

# Tilia cordata

| | |
|---|---|
| Family: | Tiliaceae |
| Scientific name: | Tilia cordata |
| English name: | Lime tree |
| German name: | Lindenbaum |
| French name: | Tilleul |
| Dutch name: | Linde |
| Part Used: | flores |
| Extraction: | watery |
| Titer: | 1/5 |
| Standard minimal: | 0,06% tilirosid, 1,2% flavonoids |
| Comment: | |

| Sorted by element name | | | Sorted by relative difference | | |
|---|---|---|---|---|---|
| aluminium | 1,8 | -0,4 | cobalt | 1,1 | 2,0 |
| arsenicum | 1,1 | 0,1 | arsenicum | 1,1 | 0,1 |
| cadmium | 0,0 | -0,3 | chromium | 0,2 | -0,2 |
| calcium | 872,8 | -0,5 | cadmium | 0,0 | -0,3 |
| chromium | 0,2 | -0,2 | copper | 2,8 | -0,3 |
| cobalt | 1,1 | 2,0 | sodium | 979,4 | -0,3 |
| copper | 2,8 | -0,3 | potassium | 17680,0 | -0,3 |
| germanium | 0,0 | -0,5 | molybdenum | 0,0 | -0,4 |
| iron | 4,4 | -0,4 | selenium | 0,0 | -0,4 |
| lead | 0,0 | -0,4 | manganese | 5,6 | -0,4 |
| lithium | 0,0 | -0,5 | vanadium | 0,2 | -0,4 |
| magnesium | 1080,0 | -0,6 | aluminium | 1,8 | -0,4 |
| manganese | 5,6 | -0,4 | zinc | 8,9 | -0,4 |
| mercury | 0,0 | -0,4 | iron | 4,4 | -0,4 |
| molybdenum | 0,0 | -0,4 | lead | 0,0 | -0,4 |
| nickel | 0,0 | -0,6 | mercury | 0,0 | -0,4 |
| phosphorus | 803,0 | -0,5 | lithium | 0,0 | -0,5 |
| potassium | 17680,0 | -0,3 | calcium | 872,8 | -0,5 |
| selenium | 0,0 | -0,4 | germanium | 0,0 | -0,5 |
| sodium | 979,4 | -0,3 | phosphorus | 803,0 | -0,5 |
| vanadium | 0,2 | -0,4 | magnesium | 1080,0 | -0,6 |
| zinc | 8,9 | -0,4 | nickel | 0,0 | -0,6 |

# Urtica dioica

| | |
|---|---|
| Family: | Urticaceae |
| Scientific name: | Urtica dioica |
| English name: | Nettle |
| German name: | Grosse Brennessel |
| French name: | Ortie |
| Dutch name: | Brandnetel grote |
| Part Used: | cortex |
| Extraction: | watery |
| Titer: | 1/4 |
| Standard minimal: | |
| Comment: | |

| Sorted by element name | | | Sorted by relative difference | | |
|---|---|---|---|---|---|
| aluminium | 5,7 | -0,4 | calcium | 21140,0 | 3,6 |
| arsenicum | 5,3 | 2,3 | arsenicum | 5,3 | 2,3 |
| cadmium | 0,0 | -0,3 | magnesium | 6500,0 | 2,2 |
| calcium | 21140,0 | 3,6 | potassium | 76260,0 | 2,1 |
| chromium | 0,4 | 0,3 | vanadium | 1,5 | 1,5 |
| cobalt | 0,4 | 0,2 | zinc | 38,6 | 1,3 |
| copper | 0,5 | -0,4 | chromium | 0,4 | 0,3 |
| germanium | 0,0 | -0,5 | phosphorus | 1859,0 | 0,2 |
| iron | 32,5 | -0,3 | cobalt | 0,4 | 0,2 |
| lead | 0,5 | 0,1 | lead | 0,5 | 0,1 |
| lithium | 0,0 | -0,5 | sodium | 1755,0 | -0,1 |
| magnesium | 6500,0 | 2,2 | mercury | 0,1 | -0,2 |
| manganese | 27,0 | -0,3 | iron | 32,5 | -0,3 |
| mercury | 0,1 | -0,2 | manganese | 27,0 | -0,3 |
| molybdenum | 0,0 | -0,4 | cadmium | 0,0 | -0,3 |
| nickel | 1,4 | -0,3 | nickel | 1,4 | -0,3 |
| phosphorus | 1859,0 | 0,2 | molybdenum | 0,0 | -0,4 |
| potassium | 76260,0 | 2,1 | aluminium | 5,7 | -0,4 |
| selenium | 0,0 | -0,4 | selenium | 0,0 | -0,4 |
| sodium | 1755,0 | -0,1 | copper | 0,5 | -0,4 |
| vanadium | 1,5 | 1,5 | lithium | 0,0 | -0,5 |
| zinc | 38,6 | 1,3 | germanium | 0,0 | -0,5 |

# Uva ursi

| | |
|---|---|
| Family: | Ericaceae |
| Scientific name: | Arctostaphylos uva ursi |
| English name: | Bearbarry |
| German name: | Bärentraube |
| French name: | Busserole |
| Dutch name: | Beredruif |
| Part Used: | folia |
| Extraction: | watery |
| Titer: | 1/4 |
| Standard minimal: | 10% arbutine |
| Comment: | |

| Sorted by element name | | | Sorted by relative difference | | |
|---|---|---|---|---|---|
| aluminium | 13,4 | -0,3 | arsenicum | 2,0 | 0,6 |
| arsenicum | 2,0 | 0,6 | selenium | 1,8 | 0,2 |
| cadmium | 0,0 | -0,3 | germanium | 0,3 | -0,1 |
| calcium | 2204,0 | -0,2 | nickel | 2,0 | -0,1 |
| chromium | 0,1 | -0,4 | calcium | 2204,0 | -0,2 |
| cobalt | 0,0 | -0,9 | aluminium | 13,4 | -0,3 |
| copper | 1,1 | -0,4 | cadmium | 0,0 | -0,3 |
| germanium | 0,3 | -0,1 | iron | 21,5 | -0,3 |
| iron | 21,5 | -0,3 | manganese | 12,8 | -0,4 |
| lead | 0,0 | -0,4 | chromium | 0,1 | -0,4 |
| lithium | 0,0 | -0,5 | molybdenum | 0,0 | -0,4 |
| magnesium | 999,6 | -0,6 | copper | 1,1 | -0,4 |
| manganese | 12,8 | -0,4 | zinc | 9,5 | -0,4 |
| mercury | 0,0 | -0,4 | sodium | 581,3 | -0,4 |
| molybdenum | 0,0 | -0,4 | lead | 0,0 | -0,4 |
| nickel | 2,0 | -0,1 | mercury | 0,0 | -0,4 |
| phosphorus | 877,9 | -0,5 | lithium | 0,0 | -0,5 |
| potassium | 10710,0 | -0,6 | phosphorus | 877,9 | -0,5 |
| selenium | 1,8 | 0,2 | magnesium | 999,6 | -0,6 |
| sodium | 581,3 | -0,4 | potassium | 10710,0 | -0,6 |
| vanadium | 0,0 | -0,7 | vanadium | 0,0 | -0,7 |
| zinc | 9,5 | -0,4 | cobalt | 0,0 | -0,9 |

# Vaccinium myrtillus

| | |
|---|---|
| Family: | Ericaceae |
| Scientific name: | Vaccinium myrtillus |
| English name: | Bilberry |
| German name: | |
| French name: | Myrtille |
| Dutch name: | Blauwe bosbes |
| Part Used: | herba |
| Extraction: | watery |
| Titer: | 1/3,5 |
| Standard minimal: | |
| Comment: | |

| Sorted by element name | | | Sorted by relative difference | | |
|---|---|---|---|---|---|
| aluminium | 5,5 | -0,4 | mercury | 0,8 | 1,2 |
| arsenicum | 0,0 | -0,4 | lead | 1,2 | 0,8 |
| cadmium | 0,2 | 0,4 | cobalt | 0,5 | 0,4 |
| calcium | 535,8 | -0,5 | cadmium | 0,2 | 0,4 |
| chromium | 0,2 | -0,2 | germanium | 0,4 | 0,0 |
| cobalt | 0,5 | 0,4 | chromium | 0,2 | -0,2 |
| copper | 1,0 | -0,4 | lithium | 0,5 | -0,2 |
| germanium | 0,4 | 0,0 | manganese | 36,8 | -0,2 |
| iron | 19,0 | -0,3 | iron | 19,0 | -0,3 |
| lead | 1,2 | 0,8 | molybdenum | 0,0 | -0,4 |
| lithium | 0,5 | -0,2 | aluminium | 5,5 | -0,4 |
| magnesium | 230,8 | -1,0 | copper | 1,0 | -0,4 |
| manganese | 36,8 | -0,2 | selenium | 0,0 | -0,4 |
| mercury | 0,8 | 1,2 | arsenicum | 0,0 | -0,4 |
| molybdenum | 0,0 | -0,4 | sodium | 332,8 | -0,5 |
| nickel | 0,0 | -0,6 | calcium | 535,8 | -0,5 |
| phosphorus | 263,6 | -0,9 | nickel | 0,0 | -0,6 |
| potassium | 2114,0 | -1,0 | zinc | 5,5 | -0,6 |
| selenium | 0,0 | -0,4 | vanadium | 0,0 | -0,7 |
| sodium | 332,8 | -0,5 | phosphorus | 263,6 | -0,9 |
| vanadium | 0,0 | -0,7 | potassium | 2114,0 | -1,0 |
| zinc | 5,5 | -0,6 | magnesium | 230,8 | -1,0 |

# Valeriana officinalis

| | |
|---|---|
| Family: | Valerianaceae |
| Scientific name: | Valeriana officinalis |
| English name: | Valerian |
| German name: | Baldrian |
| French name: | Herbe aux chats |
| Dutch name: | Valeriaan |
| Part Used: | radix |
| Extraction: | watery |
| Titer: | 1/4 |
| Standard minimal: | 0,15% cyclopentansesquiterpenes |
| Comment: | |

| Sorted by element name | | | Sorted by relative difference | | |
|---|---|---|---|---|---|
| aluminium | 34,2 | 0,1 | molybdenum | 1,0 | 1,1 |
| arsenicum | 1,4 | 0,3 | arsenicum | 1,4 | 0,3 |
| cadmium | 0,1 | 0,0 | aluminium | 34,2 | 0,1 |
| calcium | 202,3 | -0,6 | lithium | 0,8 | 0,0 |
| chromium | 0,1 | -0,4 | cadmium | 0,1 | 0,0 |
| cobalt | 0,1 | -0,6 | iron | 57,3 | -0,1 |
| copper | 2,1 | -0,3 | sodium | 1007,0 | -0,3 |
| germanium | 0,0 | -0,5 | copper | 2,1 | -0,3 |
| iron | 57,3 | -0,1 | chromium | 0,1 | -0,4 |
| lead | 0,0 | -0,4 | manganese | 8,3 | -0,4 |
| lithium | 0,8 | 0,0 | selenium | 0,0 | -0,4 |
| magnesium | 246,4 | -1,0 | lead | 0,0 | -0,4 |
| manganese | 8,3 | -0,4 | mercury | 0,0 | -0,4 |
| mercury | 0,0 | -0,4 | germanium | 0,0 | -0,5 |
| molybdenum | 1,0 | 1,1 | cobalt | 0,1 | -0,6 |
| nickel | 0,0 | -0,6 | calcium | 202,3 | -0,6 |
| phosphorus | 406,0 | -0,8 | nickel | 0,0 | -0,6 |
| potassium | 3467,0 | -0,9 | vanadium | 0,0 | -0,7 |
| selenium | 0,0 | -0,4 | phosphorus | 406,0 | -0,8 |
| sodium | 1007,0 | -0,3 | zinc | 1,5 | -0,9 |
| vanadium | 0,0 | -0,7 | potassium | 3467,0 | -0,9 |
| zinc | 1,5 | -0,9 | magnesium | 246,4 | -1,0 |

# Vinca minor

| | |
|---|---|
| Family: | Apocynaceae |
| Scientific name: | Vinca minor |
| English name: | Common periwinkle |
| German name: | Kleines Immergrün |
| French name: | Petite pervenche |
| Dutch name: | Maagdepalm |
| Part Used: | herba |
| Extraction: | alcoholic watery |
| Titer: | 1/4 |
| Standard minimal: | 0,05% vincamin |
| Comment: | |

| Sorted by element name | | | Sorted by relative difference | | |
|---|---|---|---|---|---|
| aluminium | 33,7 | 0,1 | cobalt | 0,9 | 1,5 |
| arsenicum | 0,0 | -0,4 | nickel | 6,1 | 0,9 |
| cadmium | 0,0 | -0,3 | manganese | 183,7 | 0,8 |
| calcium | 1423,0 | -0,4 | chromium | 0,3 | 0,2 |
| chromium | 0,3 | 0,2 | potassium | 28120,0 | 0,1 |
| cobalt | 0,9 | 1,5 | zinc | 17,8 | 0,1 |
| copper | 7,1 | 0,1 | copper | 7,1 | 0,1 |
| germanium | 0,0 | -0,5 | aluminium | 33,7 | 0,1 |
| iron | 29,7 | -0,3 | phosphorus | 1522,0 | 0,0 |
| lead | 0,0 | -0,4 | sodium | 1206,0 | -0,2 |
| lithium | 0,0 | -0,5 | cadmium | 0,0 | -0,3 |
| magnesium | 1595,0 | -0,3 | iron | 29,7 | -0,3 |
| manganese | 183,7 | 0,8 | vanadium | 0,3 | -0,3 |
| mercury | 0,0 | -0,4 | magnesium | 1595,0 | -0,3 |
| molybdenum | 0,0 | -0,4 | molybdenum | 0,0 | -0,4 |
| nickel | 6,1 | 0,9 | calcium | 1423,0 | -0,4 |
| phosphorus | 1522,0 | 0,0 | selenium | 0,0 | -0,4 |
| potassium | 28120,0 | 0,1 | arsenicum | 0,0 | -0,4 |
| selenium | 0,0 | -0,4 | lead | 0,0 | -0,4 |
| sodium | 1206,0 | -0,2 | mercury | 0,0 | -0,4 |
| vanadium | 0,3 | -0,3 | lithium | 0,0 | -0,5 |
| zinc | 17,8 | 0,1 | germanium | 0,0 | -0,5 |

# Viola tricolor

| | |
|---|---|
| Family: | Violaceae |
| Scientific name: | Viola tricolor |
| English name: | Pansy |
| German name: | Feldstiefmütterchen |
| French name: | Pensée des champs |
| Dutch name: | Viooltje driekleurig |
| Part Used: | herba |
| Extraction: | watery |
| Titer: | 1/6 |
| Standard minimal: | |
| Comment: | |

| Sorted by element name | | | Sorted by relative difference | | |
|---|---|---|---|---|---|
| aluminium | 149,2 | 1,8 | phosphorus | 8758,0 | 4,9 |
| arsenicum | 0,0 | -0,4 | magnesium | 6278,0 | 2,1 |
| cadmium | 0,1 | 0,1 | aluminium | 149,2 | 1,8 |
| calcium | 5017,0 | 0,4 | iron | 305,1 | 1,6 |
| chromium | 0,3 | 0,1 | zinc | 40,0 | 1,3 |
| cobalt | 0,0 | -0,9 | manganese | 190,7 | 0,9 |
| copper | 1,7 | -0,3 | lead | 1,2 | 0,8 |
| germanium | 0,0 | -0,5 | vanadium | 1,0 | 0,8 |
| iron | 305,1 | 1,6 | calcium | 5017,0 | 0,4 |
| lead | 1,2 | 0,8 | cadmium | 0,1 | 0,1 |
| lithium | 0,0 | -0,5 | chromium | 0,3 | 0,1 |
| magnesium | 6278,0 | 2,1 | copper | 1,7 | -0,3 |
| manganese | 190,7 | 0,9 | sodium | 860,5 | -0,3 |
| mercury | 0,0 | -0,4 | molybdenum | 0,0 | -0,4 |
| molybdenum | 0,0 | -0,4 | selenium | 0,0 | -0,4 |
| nickel | 0,7 | -0,5 | arsenicum | 0,0 | -0,4 |
| phosphorus | 8758,0 | 4,9 | mercury | 0,0 | -0,4 |
| potassium | 1033,0 | -1,0 | lithium | 0,0 | -0,5 |
| selenium | 0,0 | -0,4 | nickel | 0,7 | -0,5 |
| sodium | 860,5 | -0,3 | germanium | 0,0 | -0,5 |
| vanadium | 1,0 | 0,8 | cobalt | 0,0 | -0,9 |
| zinc | 40,0 | 1,3 | potassium | 1033,0 | -1,0 |

# Viscum album

| | |
|---|---|
| Family: | Loranthaceae |
| Scientific name: | Viscum album |
| English name: | Mistletoe |
| German name: | Mistel |
| French name: | Gui commun |
| Dutch name: | Maretak, vogellijm |
| Part Used: | herba |
| Extraction: | watery |
| Titer: | 1/4 |
| Standard minimal: | |
| Comment: | |

| Sorted by element name | | | Sorted by relative difference | | |
|---|---|---|---|---|---|
| aluminium | 293,1 | 4,0 | nickel | 20,5 | 4,5 |
| arsenicum | 0,0 | -0,4 | aluminium | 293,1 | 4,0 |
| cadmium | 0,0 | -0,3 | phosphorus | 4431,0 | 2,0 |
| calcium | 1107,0 | -0,4 | manganese | 273,0 | 1,4 |
| chromium | 0,5 | 0,5 | potassium | 55160,0 | 1,3 |
| cobalt | 0,6 | 0,7 | magnesium | 3721,0 | 0,8 |
| copper | 1,2 | -0,4 | cobalt | 0,6 | 0,7 |
| germanium | 0,0 | -0,5 | zinc | 28,3 | 0,7 |
| iron | 88,8 | 0,1 | chromium | 0,5 | 0,5 |
| lead | 0,0 | -0,4 | iron | 88,8 | 0,1 |
| lithium | 0,0 | -0,5 | cadmium | 0,0 | -0,3 |
| magnesium | 3721,0 | 0,8 | molybdenum | 0,0 | -0,4 |
| manganese | 273,0 | 1,4 | copper | 1,2 | -0,4 |
| mercury | 0,0 | -0,4 | selenium | 0,0 | -0,4 |
| molybdenum | 0,0 | -0,4 | calcium | 1107,0 | -0,4 |
| nickel | 20,5 | 4,5 | arsenicum | 0,0 | -0,4 |
| phosphorus | 4431,0 | 2,0 | lead | 0,0 | -0,4 |
| potassium | 55160,0 | 1,3 | mercury | 0,0 | -0,4 |
| selenium | 0,0 | -0,4 | sodium | 479,9 | -0,5 |
| sodium | 479,9 | -0,5 | lithium | 0,0 | -0,5 |
| vanadium | 0,0 | -0,7 | germanium | 0,0 | -0,5 |
| zinc | 28,3 | 0,7 | vanadium | 0,0 | -0,7 |

# Vitis vinifera

| | |
|---|---|
| Family: | Verbenaceae |
| Scientific name: | Vitis vinifera |
| English name: | Vine tree |
| German name: | Wein |
| French name: | Vigne |
| Dutch name: | Wijn |
| Part Used: | folia |
| Extraction: | watery |
| Titer: | 1/5 |
| Standard minimal: | |
| Comment: | |

| Sorted by element name | | | Sorted by relative difference | | |
|---|---|---|---|---|---|
| aluminium | 112,3 | 1,3 | copper | | 8,1 |
| arsenicum | 0,0 | -0,4 | selenium | 19,4 | 5,5 |
| cadmium | 0,1 | 0,1 | molybdenum | 4,0 | 5,4 |
| calcium | 14210,0 | 2,2 | magnesium | 10910,0 | 4,4 |
| chromium | 1,0 | 1,6 | nickel | 15,9 | 3,4 |
| cobalt | 0,3 | -0,1 | calcium | 14210,0 | 2,2 |
| copper | | 8,1 | chromium | 1,0 | 1,6 |
| germanium | 0,0 | -0,5 | iron | 277,6 | 1,4 |
| iron | 277,6 | 1,4 | manganese | 268,3 | 1,4 |
| lead | 0,0 | -0,4 | aluminium | 112,3 | 1,3 |
| lithium | 1,2 | 0,3 | zinc | 32,8 | 0,9 |
| magnesium | 10910,0 | 4,4 | vanadium | 0,8 | 0,5 |
| manganese | 268,3 | 1,4 | lithium | 1,2 | 0,3 |
| mercury | 0,0 | -0,4 | phosphorus | 1706,0 | 0,1 |
| molybdenum | 4,0 | 5,4 | cadmium | 0,1 | 0,1 |
| nickel | 15,9 | 3,4 | cobalt | 0,3 | -0,1 |
| phosphorus | 1706,0 | 0,1 | sodium | 846,0 | -0,3 |
| potassium | 10160,0 | -0,6 | arsenicum | 0,0 | -0,4 |
| selenium | 19,4 | 5,5 | lead | 0,0 | -0,4 |
| sodium | 846,0 | -0,3 | mercury | 0,0 | -0,4 |
| vanadium | 0,8 | 0,5 | germanium | 0,0 | -0,5 |
| zinc | 32,8 | 0,9 | potassium | 10160,0 | -0,6 |

# Zingiber officinalis

| | |
|---|---|
| Family: | Zingiberaceae |
| Scientific name: | Zingiber officinalis |
| English name: | Ginger |
| German name: | Ingwer |
| French name: | Gingroubre |
| Dutch name: | Gember |
| Part Used: | rhizoma |
| Extraction: | alcoholic watery |
| Titer: | 1/10 |
| Standard minimal: | |
| Comment: | |

| Sorted by element name | | | Sorted by relative difference | | |
|---|---|---|---|---|---|
| aluminium | 50,6 | 0,3 | zinc | 50,2 | 1,9 |
| arsenicum | 0,0 | -0,4 | phosphorus | 4130,0 | 1,8 |
| cadmium | 0,0 | -0,3 | potassium | 64020,0 | 1,6 |
| calcium | 1749,0 | -0,3 | copper | 23,6 | 1,3 |
| chromium | 0,6 | 0,7 | manganese | 217,0 | 1,0 |
| cobalt | 0,3 | -0,1 | nickel | 6,5 | 1,0 |
| copper | 23,6 | 1,3 | magnesium | 3840,0 | 0,8 |
| germanium | 0,0 | -0,5 | chromium | 0,6 | 0,7 |
| iron | 78,6 | 0,1 | aluminium | 50,6 | 0,3 |
| lead | 0,4 | 0,0 | iron | 78,6 | 0,1 |
| lithium | 0,4 | -0,2 | lead | 0,4 | 0,0 |
| magnesium | 3840,0 | 0,8 | cobalt | 0,3 | -0,1 |
| manganese | 217,0 | 1,0 | sodium | 1436,0 | -0,2 |
| mercury | 0,0 | -0,4 | lithium | 0,4 | -0,2 |
| molybdenum | 0,0 | -0,4 | cadmium | 0,0 | -0,3 |
| nickel | 6,5 | 1,0 | calcium | 1749,0 | -0,3 |
| phosphorus | 4130,0 | 1,8 | molybdenum | 0,0 | -0,4 |
| potassium | 64020,0 | 1,6 | selenium | 0,0 | -0,4 |
| selenium | 0,0 | -0,4 | arsenicum | 0,0 | -0,4 |
| sodium | 1436,0 | -0,2 | mercury | 0,0 | -0,4 |
| vanadium | 0,1 | -0,6 | germanium | 0,0 | -0,5 |
| zinc | 50,2 | 1,9 | vanadium | 0,1 | -0,6 |

# Part 2

# Elements

# Aluminium

| | | | | | |
|---|---|---|---|---|---|
| acorus calamus | 10,6 | -0,3 | hyoscyamus niger | 12,5 | -0,3 |
| aesculus hippocastanum | 5,7 | -0,4 | hypericum perforatum | 1,2 | -0,4 |
| agnus castus | 28,0 | 0,0 | juniperus communis | 4,7 | -0,4 |
| alchemilla vulgaris | 4,7 | -0,4 | lespedeza capitata | 8,9 | -0,3 |
| allium sativum | 6,2 | -0,4 | lithospermum | 18,3 | -0,2 |
| althea officinalis | 0,7 | -0,5 | lotus corniculatus | 25,8 | -0,1 |
| ananassa comosus | 5,0 | -0,4 | marrubium vulgare | 12,0 | -0,3 |
| angelica archangelica | 18,4 | -0,2 | medicago sativa | 6,9 | -0,4 |
| arctium lappa | 26,4 | -0,1 | melilotus officinalis | 3,3 | -0,4 |
| ballota foetida | 18,5 | -0,2 | melissa officinalis | 7,2 | -0,4 |
| berberis vulgaris | 6,9 | -0,4 | mentha piperita | 6,6 | -0,4 |
| beta vulgaris | 5,5 | -0,4 | millefolium | 0,0 | -0,5 |
| betula alba | 15,8 | -0,2 | nasturtium officinalis | 55,4 | 0,4 |
| boldo fragrans | 4,2 | -0,4 | olea europaea | 5,6 | -0,4 |
| carduus marianus | 1,9 | -0,4 | orthosiphon stamineus | 2,9 | -0,4 |
| carica papaya | 4,3 | -0,4 | passiflora incarnata | 8,4 | -0,3 |
| caroube | 450,8 | 6,4 | phaseolus vulgaris | 4,5 | -0,4 |
| carragheen | 20,1 | -0,2 | pinus maritima | 166,4 | 2,1 |
| carum carvi | 8,3 | -0,3 | piper methysticum | 0,0 | -0,5 |
| chamomilla | 12,0 | -0,3 | plantago lanceolata | 13,1 | -0,3 |
| china officinalis | 8,4 | -0,3 | raphanus sativus niger | 4,9 | -0,4 |
| chrysanthellum amer | 27,8 | 0,0 | rhamnus frangula | 1,0 | -0,5 |
| chrysanthemum parth | 8,9 | -0,3 | rhamnus purshiana | 8,6 | -0,3 |
| crataegus oxyacantha | 0,0 | -0,5 | rheum officinale | 15,8 | -0,2 |
| cupressus semp | 105,4 | 1,1 | ribes nigrum folia | 13,7 | -0,3 |
| curcuma | 96,0 | 1,0 | ribes nigrum fructus | 10,2 | -0,3 |
| cynara scolymus | 7,7 | -0,3 | rosa canina | 1,8 | -0,4 |
| echinacea angustifolia | 5,5 | -0,4 | rosmarinus officinalis | 15,8 | -0,2 |
| echinacea purpurea | 5,3 | -0,4 | ruscus aculeatus | 2,3 | -0,4 |
| eleutherococcus | 6,4 | -0,4 | salix alba | 32,4 | 0,0 |
| equisetum arvense | 4,7 | -0,4 | salvia officinalis | 18,7 | -0,2 |
| erigeron canadensis | 0,0 | -0,5 | sarsaparilla | 7,3 | -0,4 |
| eschscholtzia cali | 125,6 | 1,5 | senna | 4,9 | -0,4 |
| eucalyptus | 8,6 | -0,3 | solidago virga aurea | 3,1 | -0,4 |
| eugenia caryophyllata | 0,0 | -0,5 | spiraea ulmaria | 0,2 | -0,5 |
| fenugrec | 6,9 | -0,4 | spiruline nebl | 6,6 | -0,4 |
| foeniculum vulgare | 21,3 | -0,1 | spiruline poud | 9,0 | -0,3 |
| fraxinus excelior | 39,0 | 0,1 | taraxacum | 11,5 | -0,3 |
| fucus vesiculosus | 8,1 | -0,3 | thymus vulgaris | 20,3 | -0,2 |
| fumaria officinalis | 78,0 | 0,7 | tilia alburnum | 31,5 | 0,0 |
| gentiana lutea | 110,7 | 1,2 | tilia cordata | 1,8 | -0,4 |
| ginkgo biloba | 13,4 | -0,3 | urtica dioica | 5,7 | -0,4 |
| ginseng | 2,9 | -0,4 | uva ursi | 13,4 | -0,3 |
| glycyrrhiza glabra | 19,0 | -0,2 | vaccinium myrtillus | 5,5 | -0,4 |
| hamamelis virginia | 2,5 | -0,4 | valeriana officinalis | 34,2 | 0,1 |
| harpagophytum proc. | 10,4 | -0,3 | vinca minor | 33,7 | 0,1 |
| hibiscus sabdariffa | 51,3 | 0,3 | viola tricolor | 149,2 | 1,8 |
| hieracium pilosella | 296,5 | 4,1 | viscum album | 293,1 | 4,0 |
| humulus lupulus | 1,1 | -0,4 | vitis vinifera | 112,3 | 1,3 |
| hydrocotyle asiatica | 14,2 | -0,2 | zingiber officinalis | 50,6 | 0,3 |

# Aluminium

| | | | | |
|---|---|---|---|---|
| caroube | 450,8 | 6,4 | eucalyptus | 8,6 | -0,3 |
| hieracium pilosella | 296,5 | 4,1 | passiflora incarnata | 8,4 | -0,3 |
| viscum album | 293,1 | 4,0 | china officinalis | 8,4 | -0,3 |
| pinus maritima | 166,4 | 2,1 | rhamnus purshiana | 8,6 | -0,3 |
| viola tricolor | 149,2 | 1,8 | cynara scolymus | 7,7 | -0,3 |
| eschscholtzia cali | 125,6 | 1,5 | fucus vesiculosus | 8,1 | -0,3 |
| vitis vinifera | 112,3 | 1,3 | melissa officinalis | 7,2 | -0,4 |
| gentiana lutea | 110,7 | 1,2 | sarsaparilla | 7,3 | -0,4 |
| cupressus sem | 105,4 | 1,1 | berberis vulgaris | 6,9 | -0,4 |
| curcuma | 96,0 | 1,0 | eleutherococcus | 6,4 | -0,4 |
| fumaria officinalis | 78,0 | 0,7 | fenugrec | 6,9 | -0,4 |
| nasturtium officinalis | 55,4 | 0,4 | medicago sativa | 6,9 | -0,4 |
| hibiscus sabdariffa | 51,3 | 0,3 | mentha piperita | 6,6 | -0,4 |
| zingiber officinalis | 50,6 | 0,3 | spiruline nebl | 6,6 | -0,4 |
| fraxinus excelior | 39,0 | 0,1 | aesculus hippocastanum | 5,7 | -0,4 |
| valeriana officinalis | 34,2 | 0,1 | allium sativum | 6,2 | -0,4 |
| vinca minor | 33,7 | 0,1 | olea europaea | 5,6 | -0,4 |
| salix alba | 32,4 | 0,0 | urtica dioica | 5,7 | -0,4 |
| tilia alburnum | 31,5 | 0,0 | ananassa comosus | 5,0 | -0,4 |
| agnus castus | 28,0 | 0,0 | beta vulgaris | 5,5 | -0,4 |
| chrysanthellum amer | 27,8 | 0,0 | echinacea angustifolia | 5,5 | -0,4 |
| arctium lappa | 26,4 | -0,1 | echinacea purpurea | 5,3 | -0,4 |
| lotus corniculatus | 25,8 | -0,1 | vaccinium myrtillus | 5,5 | -0,4 |
| foeniculum vulgare | 21,3 | -0,1 | alchemilla vulgaris | 4,7 | -0,4 |
| carragheen | 20,1 | -0,2 | senna | 4,9 | -0,4 |
| thymus vulgaris | 20,3 | -0,2 | equisetum arvense | 4,7 | -0,4 |
| glycyrrhiza glabra | 19,0 | -0,2 | juniperus communis | 4,7 | -0,4 |
| angelica archangelica | 18,4 | -0,2 | carica papaya | 4,3 | -0,4 |
| ballota foetida | 18,5 | -0,2 | phaseolus vulgaris | 4,5 | -0,4 |
| lithospermum | 18,3 | -0,2 | raphanus sativus niger | 4,9 | -0,4 |
| salvia officinalis | 18,7 | -0,2 | boldo fragrans | 4,2 | -0,4 |
| betula alba | 15,8 | -0,2 | melilotus officinalis | 3,3 | -0,4 |
| rheum officinale | 15,8 | -0,2 | solidago virga aurea | 3,1 | -0,4 |
| rosmarinus officinalis | 15,8 | -0,2 | hamamelis virginia | 2,5 | -0,4 |
| hydrocotyle asiatica | 14,2 | -0,2 | orthosiphon stamineus | 2,9 | -0,4 |
| ribes nigrum folia | 13,7 | -0,3 | ginseng | 2,9 | -0,4 |
| uva ursi | 13,4 | -0,3 | carduus marianus | 1,9 | -0,4 |
| ginkgo biloba | 13,4 | -0,3 | rosa canina | 1,8 | -0,4 |
| plantago lanceolata | 13,1 | -0,3 | ruscus aculeatus | 2,3 | -0,4 |
| hyoscyamus niger | 12,5 | -0,3 | tilia cordata | 1,8 | -0,4 |
| marrubium vulgare | 12,0 | -0,3 | humulus lupulus | 1,1 | -0,4 |
| chamomilla | 12,0 | -0,3 | hypericum perforatum | 1,2 | -0,4 |
| taraxacum | 11,5 | -0,3 | althea officinalis | 0,7 | -0,5 |
| acorus calamus | 10,6 | -0,3 | rhamnus frangula | 1,0 | -0,5 |
| harpagophytum proc | 10,4 | -0,3 | millefolium | 0,0 | -0,5 |
| ribes nigrum fructus | 10,2 | -0,3 | crataegus oxyacantha | 0,0 | -0,5 |
| chrysanthemum part | 8,9 | -0,3 | erigeron canadensis | 0,0 | -0,5 |
| lespedeza capitata | 8,9 | -0,3 | eugenia caryophyllata | 0,0 | -0,5 |
| spiruline poud | 9,0 | -0,3 | piper methysticum | 0,0 | -0,5 |
| carum carvi | 8,3 | -0,3 | spiraea ulmaria | 0,2 | -0,5 |

# Arsenicum

| | | | | | |
|---|---|---|---|---|---|
| acorus calamus | 8,7 | 4,0 | hyoscyamus niger | 0,0 | -0,4 |
| aesculus hippocastanum | 0,0 | -0,4 | hypericum perforatum | 0,0 | -0,4 |
| agnus castus | 0,7 | -0,1 | juniperus communis | 0,0 | -0,4 |
| alchemilla vulgaris | 0,0 | -0,4 | lespedeza capitata | 0,5 | -0,2 |
| allium sativum | 1,4 | 0,3 | lithospermum | 0,0 | -0,4 |
| althea officinalis | 0,0 | -0,4 | lotus corniculatus | 2,8 | 1,0 |
| ananassa comosus | 3,1 | 1,1 | marrubium vulgare | 0,0 | -0,4 |
| angelica archangelica | 0,0 | -0,4 | medicago sativa | 0,0 | -0,4 |
| arctium lappa | 2,7 | 0,9 | melilotus officinalis | 0,0 | -0,4 |
| ballota foetida | 1,8 | 0,5 | melissa officinalis | 0,0 | -0,4 |
| berberis vulgaris | 0,0 | -0,4 | mentha piperita | 0,0 | -0,4 |
| beta vulgaris | 0,0 | -0,4 | millefolium | 0,0 | -0,4 |
| betula alba | 0,0 | -0,4 | nasturtium officinalis | 8,0 | 3,6 |
| boldo fragrans | 0,0 | -0,4 | olea europaea | 0,5 | -0,2 |
| carduus marianus | 1,9 | 0,5 | orthosiphon stamineus | 0,0 | -0,4 |
| carica papaya | 3,3 | 1,2 | passiflora incarnata | 2,0 | 0,6 |
| caroube | 0,0 | -0,4 | phaseolus vulgaris | 0,0 | -0,4 |
| carragheen | 0,0 | -0,4 | pinus maritima | 1,1 | 0,1 |
| carum carvi | 0,0 | -0,4 | piper methysticum | 0,0 | -0,4 |
| chamomilla | 0,0 | -0,4 | plantago lanceolata | 0,0 | -0,4 |
| china officinalis | 0,6 | -0,1 | raphanus sativus niger | 0,0 | -0,4 |
| chrysanthellum amer | 0,0 | -0,4 | rhamnus frangula | 0,7 | -0,1 |
| chrysanthemum part | 0,0 | -0,4 | rhamnus purshiana | 0,0 | -0,4 |
| crataegus oxyacantha | 0,0 | -0,4 | rheum officinale | 1,7 | 0,4 |
| cupressus semp | 0,0 | -0,4 | ribes nigrum folia | 0,0 | -0,4 |
| curcuma | 0,0 | -0,4 | ribes nigrum fructus | 0,0 | -0,4 |
| cynara scolymus | 0,9 | 0,0 | rosa canina | 0,3 | -0,3 |
| echinacea angustifolia | 0,0 | -0,4 | rosmarinus officinalis | 0,2 | -0,3 |
| echinacea purpurea | 0,0 | -0,4 | ruscus aculeatus | 0,0 | -0,4 |
| eleutherococcus | 0,0 | -0,4 | salix alba | 0,0 | -0,4 |
| equisetum arvense | 0,0 | -0,4 | salvia officinalis | 0,0 | -0,4 |
| erigeron canadensis | 0,0 | -0,4 | sarsaparilla | 0,1 | -0,4 |
| eschscholtzia cali | 2,6 | 0,9 | senna | 0,0 | -0,4 |
| eucalyptus | 0,0 | -0,4 | solidago virga aurea | 0,4 | -0,2 |
| eugenia caryophyllata | 0,0 | -0,4 | spiraea ulmaria | 0,0 | -0,4 |
| fenugrec | 0,0 | -0,4 | spiruline nebl | 0,0 | -0,4 |
| foeniculum vulgare | 0,0 | -0,4 | spiruline poud | 6,7 | 3,0 |
| fraxinus excelior | 5,1 | 2,2 | taraxacum | 0,0 | -0,4 |
| fucus vesiculosus | 9,2 | 4,3 | thymus vulgaris | 0,0 | -0,4 |
| fumaria officinalis | 0,0 | -0,4 | tilia alburnum | 0,0 | -0,4 |
| gentiana lutea | 0,0 | -0,4 | tilia cordata | 1,1 | 0,1 |
| ginkgo biloba | 0,7 | -0,1 | urtica dioica | 5,3 | 2,3 |
| ginseng | 0,0 | -0,4 | uva ursi | 2,0 | 0,6 |
| glycyrrhiza glabra | 0,4 | -0,2 | vaccinium myrtillus | 0,0 | -0,4 |
| hamamelis virginia | 0,0 | -0,4 | valeriana officinalis | 1,4 | 0,3 |
| harpagophytum proc | 0,0 | -0,4 | vinca minor | 0,0 | -0,4 |
| hibiscus sabdariffa | 0,9 | 0,0 | viola tricolor | 0,0 | -0,4 |
| hieracium pilosella | 7,8 | 3,5 | viscum album | 0,0 | -0,4 |
| humulus lupulus | 0,3 | -0,3 | vitis vinifera | 0,0 | -0,4 |
| hydrocotyle asiatica | 0,0 | -0,4 | zingiber officinalis | 0,0 | -0,4 |

# Arsenicum

| | | | | | |
|---|---|---|---|---|---|
| fucus vesiculosus | 9,2 | 4,3 | crataegus oxyacantha | 0,0 | -0,4 |
| acorus calamus | 8,7 | 4,0 | cupressus semp | 0,0 | -0,4 |
| nasturtium officinalis | 8,0 | 3,6 | curcuma | 0,0 | -0,4 |
| hieracium pilosella | 7,8 | 3,5 | echinacea angustifolia | 0,0 | -0,4 |
| spiruline poud | 6,7 | 3,0 | echinacea purpurea | 0,0 | -0,4 |
| urtica dioica | 5,3 | 2,3 | eleutherococcus | 0,0 | -0,4 |
| fraxinus excelior | 5,1 | 2,2 | equisetum arvense | 0,0 | -0,4 |
| carica papaya | 3,3 | 1,2 | erigeron canadensis | 0,0 | -0,4 |
| ananassa comosus | 3,1 | 1,1 | eucalyptus | 0,0 | -0,4 |
| lotus corniculatus | 2,8 | 1,0 | eugenia caryophyllata | 0,0 | -0,4 |
| arctium lappa | 2,7 | 0,9 | fenugrec | 0,0 | -0,4 |
| eschscholtzia cali | 2,6 | 0,9 | foeniculum vulgare | 0,0 | -0,4 |
| uva ursi | 2,0 | 0,6 | fumaria officinalis | 0,0 | -0,4 |
| passiflora incarnata | 2,0 | 0,6 | gentiana lutea | 0,0 | -0,4 |
| carduus marianus | 1,9 | 0,5 | boldo fragrans | 0,0 | -0,4 |
| ballota foetida | 1,8 | 0,5 | hamamelis virginia | 0,0 | -0,4 |
| rheum officinale | 1,7 | 0,4 | harpagophytum proc | 0,0 | -0,4 |
| allium sativum | 1,4 | 0,3 | hyoscyamus niger | 0,0 | -0,4 |
| valeriana officinalis | 1,4 | 0,3 | hypericum perforatum | 0,0 | -0,4 |
| pinus maritima | 1,1 | 0,1 | juniperus communis | 0,0 | -0,4 |
| tilia cordata | 1,1 | 0,1 | lithospermum | 0,0 | -0,4 |
| cynara scolymus | 0,9 | 0,0 | marrubium vulgare | 0,0 | -0,4 |
| hibiscus sabdariffa | 0,9 | 0,0 | chamomilla | 0,0 | -0,4 |
| ginkgo biloba | 0,7 | -0,1 | medicago sativa | 0,0 | -0,4 |
| rhamnus frangula | 0,7 | -0,1 | melilotus officinalis | 0,0 | -0,4 |
| agnus castus | 0,7 | -0,1 | melissa officinalis | 0,0 | -0,4 |
| china officinalis | 0,6 | -0,1 | mentha piperita | 0,0 | -0,4 |
| olea europaea | 0,5 | -0,2 | orthosiphon stamineus | 0,0 | -0,4 |
| lespedeza capitata | 0,5 | -0,2 | ginseng | 0,0 | -0,4 |
| glycyrrhiza glabra | 0,4 | -0,2 | phaseolus vulgaris | 0,0 | -0,4 |
| solidago virga aurea | 0,4 | -0,2 | piper methysticum | 0,0 | -0,4 |
| humulus lupulus | 0,3 | -0,3 | plantago lanceolata | 0,0 | -0,4 |
| rosa canina | 0,3 | -0,3 | raphanus sativus niger | 0,0 | -0,4 |
| rosmarinus officinalis | 0,2 | -0,3 | rhamnus purshiana | 0,0 | -0,4 |
| sarsaparilla | 0,1 | -0,4 | ribes nigrum folia | 0,0 | -0,4 |
| millefolium | 0,0 | -0,4 | ribes nigrum fructus | 0,0 | -0,4 |
| aesculus hippocastanum | 0,0 | -0,4 | ruscus aculeatus | 0,0 | -0,4 |
| alchemilla vulgaris | 0,0 | -0,4 | salix alba | 0,0 | -0,4 |
| althea officinalis | 0,0 | -0,4 | salvia officinalis | 0,0 | -0,4 |
| angelica archangelica | 0,0 | -0,4 | spiraea ulmaria | 0,0 | -0,4 |
| berberis vulgaris | 0,0 | -0,4 | spiruline nebl | 0,0 | -0,4 |
| beta vulgaris | 0,0 | -0,4 | taraxacum | 0,0 | -0,4 |
| betula alba | 0,0 | -0,4 | thymus vulgaris | 0,0 | -0,4 |
| caroube | 0,0 | -0,4 | tilia alburnum | 0,0 | -0,4 |
| carragheen | 0,0 | -0,4 | vaccinium myrtillus | 0,0 | -0,4 |
| carum carvi | 0,0 | -0,4 | vinca minor | 0,0 | -0,4 |
| senna | 0,0 | -0,4 | viola tricolor | 0,0 | -0,4 |
| hydrocotyle asiatica | 0,0 | -0,4 | viscum album | 0,0 | -0,4 |
| chrysanthellum amer | 0,0 | -0,4 | vitis vinifera | 0,0 | -0,4 |
| chrysanthemum part | 0,0 | -0,4 | zingiber officinalis | 0,0 | -0,4 |

# Cadmium

| | | | | | |
|---|---|---|---|---|---|
| acorus calamus | 0,0 | -0,3 | hyoscyamus niger | 0,0 | -0,3 |
| aesculus hippocastanum | 0,0 | -0,2 | hypericum perforatum | 0,2 | 0,4 |
| agnus castus | 0,0 | -0,3 | juniperus communis | 0,0 | -0,3 |
| alchemilla vulgaris | 0,0 | -0,3 | lespedeza capitata | 0,2 | 0,3 |
| allium sativum | 0,0 | -0,3 | lithospermum | 0,0 | -0,3 |
| althea officinalis | 0,0 | -0,3 | lotus corniculatus | 0,3 | 0,7 |
| ananassa comosus | 0,0 | -0,3 | marrubium vulgare | 0,0 | -0,3 |
| angelica archangelica | 0,1 | -0,1 | medicago sativa | 0,0 | -0,3 |
| arctium lappa | 0,0 | -0,3 | melilotus officinalis | 0,0 | -0,2 |
| ballota foetida | 0,1 | -0,1 | melissa officinalis | 0,1 | 0,2 |
| berberis vulgaris | 0,1 | 0,1 | mentha piperita | 0,1 | -0,1 |
| beta vulgaris | 0,0 | -0,1 | millefolium | 0,2 | 0,3 |
| betula alba | 0,0 | -0,3 | nasturtium officinalis | 0,0 | -0,3 |
| boldo fragrans | 0,0 | -0,3 | olea europaea | 0,0 | -0,3 |
| carduus marianus | 3,0 | 9,4 | orthosiphon stamineus | 0,3 | 0,6 |
| carica papaya | 0,0 | -0,3 | passiflora incarnata | 0,0 | -0,3 |
| caroube | 0,0 | -0,3 | phaseolus vulgaris | 0,0 | -0,3 |
| carragheen | 0,3 | 0,7 | pinus maritima | 0,0 | -0,3 |
| carum carvi | 0,1 | 0,0 | piper methysticum | 0,0 | -0,3 |
| chamomilla | 0,0 | -0,3 | plantago lanceolata | 0,2 | 0,4 |
| china officinalis | 0,0 | -0,3 | raphanus sativus niger | 0,0 | -0,3 |
| chrysanthellum amer | 0,0 | -0,3 | rhamnus frangula | 0,0 | -0,3 |
| chrysanthemum part | 0,0 | -0,3 | rhamnus purshiana | 0,2 | 0,4 |
| crataegus oxyacantha | 0,0 | -0,3 | rheum officinale | 0,0 | -0,3 |
| cupressus semp | 0,0 | -0,3 | ribes nigrum folia | 0,0 | -0,3 |
| curcuma | 0,0 | -0,3 | ribes nigrum fructus | 0,0 | -0,3 |
| cynara scolymus | 0,1 | 0,1 | rosa canina | 0,3 | 0,7 |
| echinacea angustifolia | 0,0 | -0,3 | rosmarinus officinalis | 0,0 | -0,3 |
| echinacea purpurea | 0,1 | 0,1 | ruscus aculeatus | 0,0 | -0,1 |
| eleutherococcus | 0,0 | -0,3 | salix alba | 0,0 | -0,3 |
| equisetum arvense | 0,0 | -0,3 | salvia officinalis | 0,1 | -0,1 |
| erigeron canadensis | 0,0 | -0,3 | sarsaparilla | 0,0 | -0,3 |
| eschscholtzia cali | 0,0 | -0,3 | senna | 0,3 | 0,6 |
| eucalyptus | 0,1 | 0,1 | solidago virga aurea | 0,0 | -0,3 |
| eugenia caryophyllata | 0,0 | -0,3 | spiraea ulmaria | 0,1 | 0,1 |
| fenugrec | 0,0 | -0,3 | spiruline nebl | 0,0 | -0,3 |
| foeniculum vulgare | 0,0 | -0,3 | spiruline poud | 0,3 | 0,6 |
| fraxinus excelior | 0,0 | -0,3 | taraxacum | 0,1 | 0,0 |
| fucus vesiculosus | 0,0 | -0,3 | thymus vulgaris | 0,1 | 0,0 |
| fumaria officinalis | 0,3 | 0,7 | tilia alburnum | 0,0 | -0,3 |
| gentiana lutea | 0,0 | -0,3 | tilia cordata | 0,0 | -0,3 |
| ginkgo biloba | 0,0 | -0,3 | urtica dioica | 0,0 | -0,3 |
| ginseng | 0,1 | 0,1 | uva ursi | 0,0 | -0,3 |
| glycyrrhiza glabra | 0,0 | -0,3 | vaccinium myrtillus | 0,2 | 0,4 |
| hamamelis virginia | 0,0 | -0,3 | valeriana officinalis | 0,1 | 0,0 |
| harpagophytum proc | 0,0 | -0,3 | vinca minor | 0,0 | -0,3 |
| hibiscus sabdariffa | 0,0 | -0,3 | viola tricolor | 0,1 | 0,1 |
| hieracium pilosella | 0,0 | -0,3 | viscum album | 0,0 | -0,3 |
| humulus lupulus | 0,0 | -0,3 | vitis vinifera | 0,1 | 0,1 |
| hydrocotyle asiatica | 0,0 | -0,3 | zingiber officinalis | 0,0 | -0,3 |

# Cadmium

| | | | | | |
|---|---|---|---|---|---|
| carduus marianus | 3,0 | 9,4 | echinacea angustifolia | 0,0 | -0,3 |
| carragheen | 0,3 | 0,7 | olea europaea | 0,0 | -0,3 |
| fumaria officinalis | 0,3 | 0,7 | eleutherococcus | 0,0 | -0,3 |
| lotus corniculatus | 0,3 | 0,7 | equisetum arvense | 0,0 | -0,3 |
| rosa canina | 0,3 | 0,7 | erigeron canadensis | 0,0 | -0,3 |
| senna | 0,3 | 0,6 | eschscholtzia cali | 0,0 | -0,3 |
| orthosiphon stamineus | 0,3 | 0,6 | eugenia caryophyllata | 0,0 | -0,3 |
| spiruline poud | 0,3 | 0,6 | fenugrec | 0,0 | -0,3 |
| hypericum perforatum | 0,2 | 0,4 | foeniculum vulgare | 0,0 | -0,3 |
| rhamnus purshiana | 0,2 | 0,4 | fraxinus excelior | 0,0 | -0,3 |
| vaccinium myrtillus | 0,2 | 0,4 | fucus vesiculosus | 0,0 | -0,3 |
| plantago lanceolata | 0,2 | 0,4 | gentiana lutea | 0,0 | -0,3 |
| lespedeza capitata | 0,2 | 0,3 | ginkgo biloba | 0,0 | -0,3 |
| millefolium | 0,2 | 0,3 | glycyrrhiza glabra | 0,0 | -0,3 |
| melissa officinalis | 0,1 | 0,2 | boldo fragrans | 0,0 | -0,3 |
| berberis vulgaris | 0,1 | 0,1 | hibiscus sabdariffa | 0,0 | -0,3 |
| cynara scolymus | 0,1 | 0,1 | hamamelis virginia | 0,0 | -0,3 |
| echinacea purpurea | 0,1 | 0,1 | harpagophytum proc | 0,0 | -0,3 |
| eucalyptus | 0,1 | 0,1 | hieracium pilosella | 0,0 | -0,3 |
| ginseng | 0,1 | 0,1 | humulus lupulus | 0,0 | -0,3 |
| spiraea ulmaria | 0,1 | 0,1 | hyoscyamus niger | 0,0 | -0,3 |
| viola tricolor | 0,1 | 0,1 | juniperus communis | 0,0 | -0,3 |
| vitis vinifera | 0,1 | 0,1 | lithospermum | 0,0 | -0,3 |
| valeriana officinalis | 0,1 | 0,0 | marrubium vulgare | 0,0 | -0,3 |
| carum carvi | 0,1 | 0,0 | chamomilla | 0,0 | -0,3 |
| taraxacum | 0,1 | 0,0 | medicago sativa | 0,0 | -0,3 |
| thymus vulgaris | 0,1 | 0,0 | carica papaya | 0,0 | -0,3 |
| ballota foetida | 0,1 | -0,1 | nasturtium officinalis | 0,0 | -0,3 |
| mentha piperita | 0,1 | -0,1 | passiflora incarnata | 0,0 | -0,3 |
| salvia officinalis | 0,1 | -0,1 | phaseolus vulgaris | 0,0 | -0,3 |
| angelica archangelica | 0,1 | -0,1 | pinus maritima | 0,0 | -0,3 |
| beta vulgaris | 0,0 | -0,1 | piper methysticum | 0,0 | -0,3 |
| ruscus aculeatus | 0,0 | -0,1 | china officinalis | 0,0 | -0,3 |
| aesculus hippocastanum | 0,0 | -0,2 | raphanus sativus niger | 0,0 | -0,3 |
| melilotus officinalis | 0,0 | -0,2 | rhamnus frangula | 0,0 | -0,3 |
| acorus calamus | 0,0 | -0,3 | rheum officinale | 0,0 | -0,3 |
| alchemilla vulgaris | 0,0 | -0,3 | ribes nigrum folia | 0,0 | -0,3 |
| allium sativum | 0,0 | -0,3 | ribes nigrum fructus | 0,0 | -0,3 |
| althea officinalis | 0,0 | -0,3 | rosmarinus officinalis | 0,0 | -0,3 |
| ananassa comosus | 0,0 | -0,3 | salix alba | 0,0 | -0,3 |
| arctium lappa | 0,0 | -0,3 | sarsaparilla | 0,0 | -0,3 |
| uva ursi | 0,0 | -0,3 | solidago virga aurea | 0,0 | -0,3 |
| betula alba | 0,0 | -0,3 | spiruline nebl | 0,0 | -0,3 |
| caroube | 0,0 | -0,3 | tilia alburnum | 0,0 | -0,3 |
| hydrocotyle asiatica | 0,0 | -0,3 | tilia cordata | 0,0 | -0,3 |
| chrysanthellum amer | 0,0 | -0,3 | urtica dioica | 0,0 | -0,3 |
| chrysanthemum part | 0,0 | -0,3 | vinca minor | 0,0 | -0,3 |
| crataegus oxyacantha | 0,0 | -0,3 | viscum album | 0,0 | -0,3 |
| cupressus semp | 0,0 | -0,3 | agnus castus | 0,0 | -0,3 |
| curcuma | 0,0 | -0,3 | zingiber officinalis | 0,0 | -0,3 |

# Calcium

| | | | | | |
|---|---|---|---|---|---|
| acorus calamus | 3570,0 | 0,1 | hyoscyamus niger | 352,8 | -0,6 |
| aesculus hippocastanum | 259,1 | -0,6 | hypericum perforatum | 227,2 | -0,6 |
| agnus castus | 2418,0 | -0,2 | juniperus communis | 656,7 | -0,5 |
| alchemilla vulgaris | 2562,0 | -0,1 | lespedeza capitata | 615,0 | -0,5 |
| allium sativum | 1021,0 | -0,5 | lithospermum | 8548,0 | 1,1 |
| althea officinalis | 368,2 | -0,6 | lotus corniculatus | 13460,0 | 2,0 |
| ananassa comosus | 4730,0 | 0,3 | marrubium vulgare | 1365,0 | -0,4 |
| angelica archangelica | 284,0 | -0,6 | medicago sativa | 4222,0 | 0,2 |
| arctium lappa | 1319,0 | -0,4 | melilotus officinalis | 5353,0 | 0,4 |
| ballota foetida | 1753,0 | -0,3 | melissa officinalis | 240,9 | -0,6 |
| berberis vulgaris | 1427,0 | -0,4 | mentha piperita | 3896,0 | 0,1 |
| beta vulgaris | 4362,0 | 0,2 | millefolium | 1378,0 | -0,4 |
| betula alba | 2855,0 | -0,1 | nasturtium officinalis | 28960,0 | 5,2 |
| boldo fragrans | 402,9 | -0,6 | olea europaea | 1614,0 | -0,3 |
| carduus marianus | 191,3 | -0,6 | orthosiphon stamineus | 402,9 | -0,6 |
| carica papaya | 3270,0 | 0,0 | passiflora incarnata | 3173,0 | 0,0 |
| caroube | 3502,0 | 0,1 | phaseolus vulgaris | 1163,0 | -0,4 |
| carragheen | 3302,0 | 0,0 | pinus maritima | 3877,0 | 0,1 |
| carum carvi | 816,2 | -0,5 | piper methysticum | 86,3 | -0,6 |
| chamomilla | 989,6 | -0,5 | plantago lanceolata | 4657,0 | 0,3 |
| china officinalis | 189,7 | -0,6 | raphanus sativus niger | 2324,0 | -0,2 |
| chrysanthellum amer | 2777,0 | -0,1 | rhamnus frangula | 592,5 | -0,5 |
| chrysanthemum part | 352,3 | -0,6 | rhamnus purshiana | 843,7 | -0,5 |
| crataegus oxyacantha | 1397,0 | -0,4 | rheum officinale | 680,0 | -0,5 |
| cupressus semp | 4062,0 | 0,2 | ribes nigrum folia | 3651,0 | 0,1 |
| curcuma | 974,1 | -0,5 | ribes nigrum fructus | 988,1 | -0,5 |
| cynara scolymus | 1382,0 | -0,4 | rosa canina | 746,0 | -0,5 |
| echinacea angustifolia | 717,9 | -0,5 | rosmarinus officinalis | 11690,0 | 1,7 |
| echinacea purpurea | 1980,0 | -0,3 | ruscus aculeatus | 414,8 | -0,6 |
| eleutherococcus | 654,9 | -0,5 | salix alba | 8850,0 | 1,1 |
| equisetum arvense | 1747,0 | -0,3 | salvia officinalis | 7311,0 | 0,8 |
| erigeron canadensis | 515,4 | -0,6 | sarsaparilla | 1996,0 | -0,3 |
| eschscholtzia cali | 2478,0 | -0,2 | senna | 332,2 | -0,6 |
| eucalyptus | 1005,0 | -0,5 | solidago virga aurea | 2731,0 | -0,1 |
| eugenia caryophyllata | 344,7 | -0,6 | spiraea ulmaria | 1251,0 | -0,4 |
| fenugrec | 254,0 | -0,6 | spiruline nebl | 1758,0 | -0,3 |
| foeniculum vulgare | 2134,0 | -0,2 | spiruline poud | 1607,0 | -0,3 |
| fraxinus excelior | 26480,0 | 4,7 | taraxacum | 3041,0 | 0,0 |
| fucus vesiculosus | 1426,0 | -0,4 | thymus vulgaris | 9903,0 | 1,3 |
| fumaria officinalis | 15100,0 | 2,4 | tilia alburnum | 3306,0 | 0,0 |
| gentiana lutea | 1568,0 | -0,3 | tilia cordata | 872,8 | -0,5 |
| ginkgo biloba | 4043,0 | 0,2 | urtica dioica | 21140,0 | 3,6 |
| ginseng | 178,2 | -0,6 | uva ursi | 2204,0 | -0,2 |
| glycyrrhiza glabra | 3824,0 | 0,1 | vaccinium myrtillus | 535,8 | -0,5 |
| hamamelis virginia | 1306,0 | -0,4 | valeriana officinalis | 202,3 | -0,6 |
| harpagophytum proc | 436,5 | -0,6 | vinca minor | 1423,0 | -0,4 |
| hibiscus sabdariffa | 7396,0 | 0,8 | viola tricolor | 5017,0 | 0,4 |
| hieracium pilosella | 3900,0 | 0,1 | viscum album | 1107,0 | -0,4 |
| humulus lupulus | 71,1 | -0,6 | vitis vinifera | 14210,0 | 2,2 |
| hydrocotyle asiatica | 596,5 | -0,5 | zingiber officinalis | 1749,0 | -0,3 |

# Calcium

| | | | | |
|---|---|---|---|---|
| nasturtium officinalis | 28960,0 | 5,2 | berberis vulgaris | 1427,0 | -0,4 |
| fraxinus excelior | 26480,0 | 4,7 | crataegus oxyacantha | 1397,0 | -0,4 |
| urtica dioica | 21140,0 | 3,6 | cynara scolymus | 1382,0 | -0,4 |
| fumaria officinalis | 15100,0 | 2,4 | fucus vesiculosus | 1426,0 | -0,4 |
| vitis vinifera | 14210,0 | 2,2 | vinca minor | 1423,0 | -0,4 |
| lotus corniculatus | 13460,0 | 2,0 | millefolium | 1378,0 | -0,4 |
| rosmarinus officinalis | 11690,0 | 1,7 | marrubium vulgare | 1365,0 | -0,4 |
| thymus vulgaris | 9903,0 | 1,3 | arctium lappa | 1319,0 | -0,4 |
| salix alba | 8850,0 | 1,1 | hamamelis virginia | 1306,0 | -0,4 |
| lithospermum | 8548,0 | 1,1 | spiraea ulmaria | 1251,0 | -0,4 |
| hibiscus sabdariffa | 7396,0 | 0,8 | phaseolus vulgaris | 1163,0 | -0,4 |
| salvia officinalis | 7311,0 | 0,8 | viscum album | 1107,0 | -0,4 |
| melilotus officinalis | 5353,0 | 0,4 | allium sativum | 1021,0 | -0,5 |
| viola tricolor | 5017,0 | 0,4 | eucalyptus | 1005,0 | -0,5 |
| ananassa comosus | 4730,0 | 0,3 | chamomilla | 989,6 | -0,5 |
| plantago lanceolata | 4657,0 | 0,3 | ribes nigrum fructus | 988,1 | -0,5 |
| beta vulgaris | 4362,0 | 0,2 | curcuma | 974,1 | -0,5 |
| medicago sativa | 4222,0 | 0,2 | rhamnus purshiana | 843,7 | -0,5 |
| cupressus semp | 4062,0 | 0,2 | tilia cordata | 872,8 | -0,5 |
| ginkgo biloba | 4043,0 | 0,2 | carum carvi | 816,2 | -0,5 |
| hieracium pilosella | 3900,0 | 0,1 | rosa canina | 746,0 | -0,5 |
| mentha piperita | 3896,0 | 0,1 | echinacea angustifolia | 717,9 | -0,5 |
| pinus maritima | 3877,0 | 0,1 | eleutherococcus | 654,9 | -0,5 |
| glycyrrhiza glabra | 3824,0 | 0,1 | juniperus communis | 656,7 | -0,5 |
| ribes nigrum folia | 3651,0 | 0,1 | rheum officinale | 680,0 | -0,5 |
| acorus calamus | 3570,0 | 0,1 | hydrocotyle asiatica | 596,5 | -0,5 |
| caroube | 3502,0 | 0,1 | lespedeza capitata | 615,0 | -0,5 |
| carragheen | 3302,0 | 0,0 | rhamnus frangula | 592,5 | -0,5 |
| tilia alburnum | 3306,0 | 0,0 | vaccinium myrtillus | 535,8 | -0,5 |
| carica papaya | 3270,0 | 0,0 | erigeron canadensis | 515,4 | -0,6 |
| passiflora incarnata | 3173,0 | 0,0 | harpagophytum proc | 436,5 | -0,6 |
| taraxacum | 3041,0 | 0,0 | boldo fragrans | 402,9 | -0,6 |
| betula alba | 2855,0 | -0,1 | orthosiphon stamineus | 402,9 | -0,6 |
| chrysanthellum amer | 2777,0 | -0,1 | ruscus aculeatus | 414,8 | -0,6 |
| solidago virga aurea | 2731,0 | -0,1 | althea officinalis | 368,2 | -0,6 |
| alchemilla vulgaris | 2562,0 | -0,1 | chrysanthemum part | 352,3 | -0,6 |
| eschscholtzia cali | 2478,0 | -0,2 | eugenia caryophyllata | 344,7 | -0,6 |
| agnus castus | 2418,0 | -0,2 | hyoscyamus niger | 352,8 | -0,6 |
| raphanus sativus niger | 2324,0 | -0,2 | angelica archangelica | 284,0 | -0,6 |
| uva ursi | 2204,0 | -0,2 | senna | 332,2 | -0,6 |
| foeniculum vulgare | 2134,0 | -0,2 | aesculus hippocastanum | 259,1 | -0,6 |
| sarsaparilla | 1996,0 | -0,3 | fenugrec | 254,0 | -0,6 |
| echinacea purpurea | 1980,0 | -0,3 | melissa officinalis | 240,9 | -0,6 |
| ballota foetida | 1753,0 | -0,3 | carduus marianus | 191,3 | -0,6 |
| equisetum arvense | 1747,0 | -0,3 | hypericum perforatum | 227,2 | -0,6 |
| spiruline nebl | 1758,0 | -0,3 | china officinalis | 189,7 | -0,6 |
| zingiber officinalis | 1749,0 | -0,3 | valeriana officinalis | 202,3 | -0,6 |
| olea europaea | 1614,0 | -0,3 | ginseng | 178,2 | -0,6 |
| spiruline poud | 1607,0 | -0,3 | piper methysticum | 86,3 | -0,6 |
| gentiana lutea | 1568,0 | -0,3 | humulus lupulus | 71,1 | -0,6 |

# Chromium

| | | | | | |
|---|---|---|---|---|---|
| acorus calamus | 0,4 | 0,3 | hyoscyamus niger | 0,1 | -0,4 |
| aesculus hippocastanum | 0,1 | -0,4 | hypericum perforatum | 0,3 | 0,1 |
| agnus castus | 0,2 | -0,2 | juniperus communis | 0,0 | -0,5 |
| alchemilla vulgaris | 0,1 | -0,4 | lespedeza capitata | 0,4 | 0,2 |
| allium sativum | 0,1 | -0,4 | lithospermum | 0,4 | 0,3 |
| althea officinalis | 0,2 | -0,2 | lotus corniculatus | 0,6 | 0,7 |
| ananassa comosus | 0,0 | -0,6 | marrubium vulgare | 0,3 | 0,1 |
| angelica archangelica | 0,1 | -0,5 | medicago sativa | 0,0 | -0,6 |
| arctium lappa | 0,7 | 0,9 | melilotus officinalis | 0,2 | -0,2 |
| ballota foetida | 0,3 | 0,1 | melissa officinalis | 0,0 | -0,6 |
| berberis vulgaris | 0,2 | -0,2 | mentha piperita | 0,0 | -0,6 |
| beta vulgaris | 0,0 | -0,5 | millefolium | 0,0 | -0,6 |
| betula alba | 0,4 | 0,3 | nasturtium officinalis | 0,8 | 1,1 |
| boldo fragrans | 0,1 | -0,5 | olea europaea | 0,3 | 0,1 |
| carduus marianus | 0,2 | -0,2 | orthosiphon stamineus | 0,0 | -0,6 |
| carica papaya | 0,1 | -0,5 | passiflora incarnata | 0,0 | -0,6 |
| caroube | 1,9 | 3,5 | phaseolus vulgaris | 0,2 | -0,2 |
| carragheen | 0,0 | -0,6 | pinus maritima | 0,4 | 0,3 |
| carum carvi | 0,2 | -0,2 | piper methysticum | 0,5 | 0,5 |
| chamomilla | 0,2 | -0,2 | plantago lanceolata | 0,2 | -0,2 |
| china officinalis | 3,9 | 7,7 | raphanus sativus niger | 0,0 | -0,6 |
| chrysanthellum amer | 0,3 | 0,1 | rhamnus frangula | 0,0 | -0,6 |
| chrysanthemum part | 0,1 | -0,4 | rhamnus purshiana | 0,0 | -0,6 |
| crataegus oxyacantha | 0,1 | -0,4 | rheum officinale | 0,0 | -0,6 |
| cupressus semp | 0,7 | 0,9 | ribes nigrum folia | 0,0 | -0,6 |
| curcuma | 1,3 | 2,2 | ribes nigrum fructus | 0,3 | 0,1 |
| cynara scolymus | 0,3 | 0,0 | rosa canina | 0,0 | -0,6 |
| echinacea angustifolia | 0,5 | 0,5 | rosmarinus officinalis | 0,1 | -0,4 |
| echinacea purpurea | 0,3 | 0,1 | ruscus aculeatus | 0,3 | 0,1 |
| eleutherococcus | 0,0 | -0,6 | salix alba | 0,4 | 0,2 |
| equisetum arvense | 0,3 | 0,1 | salvia officinalis | 0,1 | -0,4 |
| erigeron canadensis | 0,0 | -0,6 | sarsaparilla | 0,0 | -0,6 |
| eschscholtzia cali | 0,7 | 0,9 | senna | 0,2 | -0,2 |
| eucalyptus | 0,1 | -0,4 | solidago virga aurea | 0,0 | -0,6 |
| eugenia caryophyllata | 0,0 | -0,6 | spiraea ulmaria | 0,0 | -0,6 |
| fenugrec | 0,0 | -0,6 | spiruline nebl | 0,2 | -0,2 |
| foeniculum vulgare | 0,2 | -0,2 | spiruline poud | 0,2 | -0,3 |
| fraxinus excelior | 0,0 | -0,6 | taraxacum | 0,2 | -0,2 |
| fucus vesiculosus | 0,2 | -0,2 | thymus vulgaris | 0,4 | 0,3 |
| fumaria officinalis | 0,4 | 0,3 | tilia alburnum | 0,2 | -0,2 |
| gentiana lutea | 0,0 | -0,6 | tilia cordata | 0,2 | -0,2 |
| ginkgo biloba | 0,2 | -0,2 | urtica dioica | 0,4 | 0,3 |
| ginseng | 0,0 | -0,6 | uva ursi | 0,1 | -0,4 |
| glycyrrhiza glabra | 0,2 | -0,2 | vaccinium myrtillus | 0,2 | -0,2 |
| hamamelis virginia | 0,0 | -0,6 | valeriana officinalis | 0,1 | -0,4 |
| harpagophytum proc | 0,2 | -0,2 | vinca minor | 0,3 | 0,2 |
| hibiscus sabdariffa | 0,6 | 0,7 | viola tricolor | 0,3 | 0,1 |
| hieracium pilosella | 0,5 | 0,5 | viscum album | 0,5 | 0,5 |
| humulus lupulus | 0,0 | -0,5 | vitis vinifera | 1,0 | 1,6 |
| hydrocotyle asiatica | 0,1 | -0,5 | zingiber officinalis | 0,6 | 0,7 |

# Chromium

| | | | | | |
|---|---|---|---|---|---|
| china officinalis | 3,9 | 7,7 | tilia alburnum | 0,2 | -0,2 |
| caroube | 1,9 | 3,5 | tilia cordata | 0,2 | -0,2 |
| curcuma | 1,3 | 2,2 | vaccinium myrtillus | 0,2 | -0,2 |
| vitis vinifera | 1,0 | 1,6 | agnus castus | 0,2 | -0,2 |
| nasturtium officinalis | 0,8 | 1,1 | plantago lanceolata | 0,2 | -0,2 |
| arctium lappa | 0,7 | 0,9 | senna | 0,2 | -0,2 |
| cupressus semp | 0,7 | 0,9 | spiruline poud | 0,2 | -0,3 |
| eschscholtzia cali | 0,7 | 0,9 | aesculus hippocastanum | 0,1 | -0,4 |
| hibiscus sabdariffa | 0,6 | 0,7 | alchemilla vulgaris | 0,1 | -0,4 |
| lotus corniculatus | 0,6 | 0,7 | uva ursi | 0,1 | -0,4 |
| zingiber officinalis | 0,6 | 0,7 | chrysanthemum part | 0,1 | -0,4 |
| echinacea angustifolia | 0,5 | 0,5 | crataegus oxyacantha | 0,1 | -0,4 |
| hieracium pilosella | 0,5 | 0,5 | eucalyptus | 0,1 | -0,4 |
| piper methysticum | 0,5 | 0,5 | hyoscyamus niger | 0,1 | -0,4 |
| viscum album | 0,5 | 0,5 | salvia officinalis | 0,1 | -0,4 |
| acorus calamus | 0,4 | 0,3 | valeriana officinalis | 0,1 | -0,4 |
| betula alba | 0,4 | 0,3 | allium sativum | 0,1 | -0,4 |
| fumaria officinalis | 0,4 | 0,3 | rosmarinus officinalis | 0,1 | -0,4 |
| lithospermum | 0,4 | 0,3 | hydrocotyle asiatica | 0,1 | -0,5 |
| pinus maritima | 0,4 | 0,3 | boldo fragrans | 0,1 | -0,5 |
| thymus vulgaris | 0,4 | 0,3 | angelica archangelica | 0,1 | -0,5 |
| urtica dioica | 0,4 | 0,3 | carica papaya | 0,1 | -0,5 |
| salix alba | 0,4 | 0,2 | beta vulgaris | 0,0 | -0,5 |
| lespedeza capitata | 0,4 | 0,2 | humulus lupulus | 0,0 | -0,5 |
| vinca minor | 0,3 | 0,2 | juniperus communis | 0,0 | -0,5 |
| ballota foetida | 0,3 | 0,1 | melissa officinalis | 0,0 | -0,6 |
| chrysanthellum amer | 0,3 | 0,1 | millefolium | 0,0 | -0,6 |
| echinacea purpurea | 0,3 | 0,1 | ananassa comosus | 0,0 | -0,6 |
| olea europaea | 0,3 | 0,1 | carragheen | 0,0 | -0,6 |
| equisetum arvense | 0,3 | 0,1 | eleutherococcus | 0,0 | -0,6 |
| hypericum perforatum | 0,3 | 0,1 | erigeron canadensis | 0,0 | -0,6 |
| marrubium vulgare | 0,3 | 0,1 | eugenia caryophyllata | 0,0 | -0,6 |
| ribes nigrum fructus | 0,3 | 0,1 | fenugrec | 0,0 | -0,6 |
| ruscus aculeatus | 0,3 | 0,1 | fraxinus excelior | 0,0 | -0,6 |
| viola tricolor | 0,3 | 0,1 | gentiana lutea | 0,0 | -0,6 |
| cynara scolymus | 0,3 | 0,0 | hamamelis virginia | 0,0 | -0,6 |
| althea officinalis | 0,2 | -0,2 | medicago sativa | 0,0 | -0,6 |
| berberis vulgaris | 0,2 | -0,2 | mentha piperita | 0,0 | -0,6 |
| carduus marianus | 0,2 | -0,2 | orthosiphon stamineus | 0,0 | -0,6 |
| carum carvi | 0,2 | -0,2 | ginseng | 0,0 | -0,6 |
| foeniculum vulgare | 0,2 | -0,2 | passiflora incarnata | 0,0 | -0,6 |
| fucus vesiculosus | 0,2 | -0,2 | raphanus sativus niger | 0,0 | -0,6 |
| ginkgo biloba | 0,2 | -0,2 | rhamnus frangula | 0,0 | -0,6 |
| glycyrrhiza glabra | 0,2 | -0,2 | rhamnus purshiana | 0,0 | -0,6 |
| harpagophytum proc | 0,2 | -0,2 | rheum officinale | 0,0 | -0,6 |
| chamomilla | 0,2 | -0,2 | ribes nigrum folia | 0,0 | -0,6 |
| melilotus officinalis | 0,2 | -0,2 | rosa canina | 0,0 | -0,6 |
| phaseolus vulgaris | 0,2 | -0,2 | sarsaparilla | 0,0 | -0,6 |
| spiruline nebl | 0,2 | -0,2 | solidago virga aurea | 0,0 | -0,6 |
| taraxacum | 0,2 | -0,2 | spiraea ulmaria | 0,0 | -0,6 |

# Cobalt

| | | | | | | |
|---|---|---|---|---|---|
| acorus calamus | 0,7 | 1,0 | hyoscyamus niger | 0,2 | -0,3 |
| aesculus hippocastanum | 0,0 | -0,9 | hypericum perforatum | 0,0 | -0,9 |
| agnus castus | 0,2 | -0,3 | juniperus communis | 0,9 | 1,5 |
| alchemilla vulgaris | 0,4 | 0,2 | lespedeza capitata | 0,5 | 0,4 |
| allium sativum | 0,0 | -0,9 | lithospermum | 0,0 | -0,9 |
| althea officinalis | 0,5 | 0,4 | lotus corniculatus | 0,3 | -0,1 |
| ananassa comosus | 0,0 | -0,9 | marrubium vulgare | 0,5 | 0,4 |
| angelica archangelica | 0,2 | -0,3 | medicago sativa | 0,0 | -0,9 |
| arctium lappa | 1,0 | 1,7 | melilotus officinalis | 0,0 | -0,9 |
| ballota foetida | 0,3 | -0,1 | melissa officinalis | 0,3 | -0,1 |
| berberis vulgaris | 0,9 | 1,5 | mentha piperita | 0,5 | 0,4 |
| beta vulgaris | 0,1 | -0,6 | millefolium | 0,3 | -0,1 |
| betula alba | 0,5 | 0,4 | nasturtium officinalis | 0,3 | -0,1 |
| boldo fragrans | 0,0 | -0,9 | olea europaea | 0,4 | 0,2 |
| carduus marianus | 0,0 | -0,9 | orthosiphon stamineus | 0,0 | -0,9 |
| carica papaya | 0,0 | -0,9 | passiflora incarnata | 0,0 | -0,9 |
| caroube | 1,1 | 2,0 | phaseolus vulgaris | 0,0 | -0,9 |
| carragheen | 0,7 | 1,0 | pinus maritima | 0,3 | -0,1 |
| carum carvi | 0,6 | 0,7 | piper methysticum | 0,4 | 0,2 |
| chamomilla | 0,0 | -0,9 | plantago lanceolata | 0,2 | -0,3 |
| china officinalis | 0,3 | -0,1 | raphanus sativus niger | 0,1 | -0,7 |
| chrysanthellum amer | 0,3 | -0,1 | rhamnus frangula | 0,2 | -0,3 |
| chrysanthemum part | 1,0 | 1,7 | rhamnus purshiana | 0,3 | -0,1 |
| crataegus oxyacantha | 0,0 | -0,9 | rheum officinale | 0,0 | -0,9 |
| cupressus semp | 0,6 | 0,7 | ribes nigrum folia | 0,1 | -0,5 |
| curcuma | 0,0 | -0,9 | ribes nigrum fructus | 0,0 | -0,9 |
| cynara scolymus | 0,0 | -0,9 | rosa canina | 0,1 | -0,6 |
| echinacea angustifolia | 0,0 | -0,9 | rosmarinus officinalis | 0,0 | -0,9 |
| echinacea purpurea | 0,0 | -0,9 | ruscus aculeatus | 1,1 | 2,0 |
| eleutherococcus | 0,0 | -0,9 | salix alba | 1,3 | 2,5 |
| equisetum arvense | 0,5 | 0,4 | salvia officinalis | 0,3 | -0,1 |
| erigeron canadensis | 0,0 | -0,9 | sarsaparilla | 0,4 | 0,2 |
| eschscholtzia cali | 0,5 | 0,4 | senna | 0,2 | -0,4 |
| eucalyptus | 0,3 | -0,1 | solidago virga aurea | 0,1 | -0,6 |
| eugenia caryophyllata | 0,1 | -0,6 | spiraea ulmaria | 0,0 | -0,9 |
| fenugrec | 0,0 | -0,9 | spiruline nebl | 0,1 | -0,7 |
| foeniculum vulgare | 0,0 | -0,9 | spiruline poud | 0,0 | -0,9 |
| fraxinus excelior | 0,7 | 1,0 | taraxacum | 1,0 | 1,7 |
| fucus vesiculosus | 0,0 | -0,9 | thymus vulgaris | 0,1 | -0,6 |
| fumaria officinalis | 0,3 | -0,1 | tilia alburnum | 0,0 | -0,9 |
| gentiana lutea | 1,5 | 3,0 | tilia cordata | 1,1 | 2,0 |
| ginkgo biloba | 0,2 | -0,3 | urtica dioica | 0,4 | 0,2 |
| ginseng | 0,9 | 1,5 | uva ursi | 0,0 | -0,9 |
| glycyrrhiza glabra | 1,0 | 1,7 | vaccinium myrtillus | 0,5 | 0,4 |
| hamamelis virginia | 0,0 | -0,8 | valeriana officinalis | 0,1 | -0,6 |
| harpagophytum proc | 0,0 | -0,8 | vinca minor | 0,9 | 1,5 |
| hibiscus sabdariffa | 1,4 | 2,7 | viola tricolor | 0,0 | -0,9 |
| hieracium pilosella | 0,5 | 0,4 | viscum album | 0,6 | 0,7 |
| humulus lupulus | 0,0 | -0,9 | vitis vinifera | 0,3 | -0,1 |
| hydrocotyle asiatica | 1,4 | 2,7 | zingiber officinalis | 0,3 | -0,1 |

# Cobalt

| | | | | | |
|---|---|---|---|---|---|
| gentiana lutea | 1,5 | 3,0 | ginkgo biloba | 0,2 | -0,3 |
| hydrocotyle asiatica | 1,4 | 2,7 | hyoscyamus niger | 0,2 | -0,3 |
| hibiscus sabdariffa | 1,4 | 2,7 | plantago lanceolata | 0,2 | -0,3 |
| salix alba | 1,3 | 2,5 | rhamnus frangula | 0,2 | -0,3 |
| caroube | 1,1 | 2,0 | agnus castus | 0,2 | -0,3 |
| ruscus aculeatus | 1,1 | 2,0 | senna | 0,2 | -0,4 |
| tilia cordata | 1,1 | 2,0 | ribes nigrum folia | 0,1 | -0,5 |
| arctium lappa | 1,0 | 1,7 | beta vulgaris | 0,1 | -0,6 |
| chrysanthemum part | 1,0 | 1,7 | eugenia caryophyllata | 0,1 | -0,6 |
| glycyrrhiza glabra | 1,0 | 1,7 | rosa canina | 0,1 | -0,6 |
| taraxacum | 1,0 | 1,7 | solidago virga aurea | 0,1 | -0,6 |
| berberis vulgaris | 0,9 | 1,5 | thymus vulgaris | 0,1 | -0,6 |
| juniperus communis | 0,9 | 1,5 | valeriana officinalis | 0,1 | -0,6 |
| ginseng | 0,9 | 1,5 | raphanus sativus niger | 0,1 | -0,7 |
| vinca minor | 0,9 | 1,5 | spiruline nebl | 0,1 | -0,7 |
| acorus calamus | 0,7 | 1,0 | hamamelis virginia | 0,0 | -0,8 |
| carragheen | 0,7 | 1,0 | harpagophytum proc | 0,0 | -0,8 |
| fraxinus excelior | 0,7 | 1,0 | aesculus hippocastanum | 0,0 | -0,9 |
| carum carvi | 0,6 | 0,7 | allium sativum | 0,0 | -0,9 |
| cupressus semp | 0,6 | 0,7 | ananassa comosus | 0,0 | -0,9 |
| viscum album | 0,6 | 0,7 | uva ursi | 0,0 | -0,9 |
| althea officinalis | 0,5 | 0,4 | carduus marianus | 0,0 | -0,9 |
| betula alba | 0,5 | 0,4 | crataegus oxyacantha | 0,0 | -0,9 |
| equisetum arvense | 0,5 | 0,4 | curcuma | 0,0 | -0,9 |
| eschscholtzia cali | 0,5 | 0,4 | cynara scolymus | 0,0 | -0,9 |
| hieracium pilosella | 0,5 | 0,4 | echinacea angustifolia | 0,0 | -0,9 |
| lespedeza capitata | 0,5 | 0,4 | echinacea purpurea | 0,0 | -0,9 |
| marrubium vulgare | 0,5 | 0,4 | eleutherococcus | 0,0 | -0,9 |
| mentha piperita | 0,5 | 0,4 | erigeron canadensis | 0,0 | -0,9 |
| vaccinium myrtillus | 0,5 | 0,4 | fenugrec | 0,0 | -0,9 |
| alchemilla vulgaris | 0,4 | 0,2 | foeniculum vulgare | 0,0 | -0,9 |
| olea europaea | 0,4 | 0,2 | fucus vesiculosus | 0,0 | -0,9 |
| piper methysticum | 0,4 | 0,2 | boldo fragrans | 0,0 | -0,9 |
| sarsaparilla | 0,4 | 0,2 | humulus lupulus | 0,0 | -0,9 |
| urtica dioica | 0,4 | 0,2 | hypericum perforatum | 0,0 | -0,9 |
| millefolium | 0,3 | -0,1 | lithospermum | 0,0 | -0,9 |
| ballota foetida | 0,3 | -0,1 | chamomilla | 0,0 | -0,9 |
| chrysanthellum amer | 0,3 | -0,1 | medicago sativa | 0,0 | -0,9 |
| eucalyptus | 0,3 | -0,1 | melilotus officinalis | 0,0 | -0,9 |
| fumaria officinalis | 0,3 | -0,1 | carica papaya | 0,0 | -0,9 |
| lotus corniculatus | 0,3 | -0,1 | orthosiphon stamineus | 0,0 | -0,9 |
| melissa officinalis | 0,3 | -0,1 | passiflora incarnata | 0,0 | -0,9 |
| nasturtium officinalis | 0,3 | -0,1 | phaseolus vulgaris | 0,0 | -0,9 |
| pinus maritima | 0,3 | -0,1 | rheum officinale | 0,0 | -0,9 |
| china officinalis | 0,3 | -0,1 | ribes nigrum fructus | 0,0 | -0,9 |
| rhamnus purshiana | 0,3 | -0,1 | rosmarinus officinalis | 0,0 | -0,9 |
| salvia officinalis | 0,3 | -0,1 | spiraea ulmaria | 0,0 | -0,9 |
| vitis vinifera | 0,3 | -0,1 | spiruline poud | 0,0 | -0,9 |
| zingiber officinalis | 0,3 | -0,1 | tilia alburnum | 0,0 | -0,9 |
| angelica archangelica | 0,2 | -0,3 | viola tricolor | 0,0 | -0,9 |

# Copper

| | | | | | |
|---|--:|--:|---|--:|--:|
| acorus calamus | 3,4 | -0,2 | hyoscyamus niger | 12,3 | 0,4 |
| aesculus hippocastanum | 16,1 | 0,7 | hypericum perforatum | 3,1 | -0,2 |
| agnus castus | 3,0 | -0,2 | juniperus communis | 1,5 | -0,4 |
| alchemilla vulgaris | 6,3 | 0,0 | lespedeza capitata | 3,7 | -0,2 |
| allium sativum | 5,7 | 0,0 | lithospermum | 3,7 | -0,2 |
| althea officinalis | 0,0 | -0,5 | lotus corniculatus | 14,8 | 0,6 |
| ananassa comosus | 4,5 | -0,1 | marrubium vulgare | 2,6 | -0,3 |
| angelica archangelica | 1,2 | -0,4 | medicago sativa | 0,9 | -0,4 |
| arctium lappa | 4,4 | -0,1 | melilotus officinalis | 10,6 | 0,3 |
| ballota foetida | 4,7 | -0,1 | melissa officinalis | 2,6 | -0,3 |
| berberis vulgaris | 6,2 | 0,0 | mentha piperita | 0,6 | -0,4 |
| beta vulgaris | 5,3 | -0,1 | millefolium | 1,1 | -0,4 |
| betula alba | 1,2 | -0,4 | nasturtium officinalis | 1,2 | -0,4 |
| boldo fragrans | 0,5 | -0,4 | olea europaea | 3,5 | -0,2 |
| carduus marianus | 1,9 | -0,3 | orthosiphon stamineus | 27,2 | 1,5 |
| carica papaya | 1,1 | -0,4 | passiflora incarnata | 2,7 | -0,3 |
| caroube | 5,6 | -0,1 | phaseolus vulgaris | 22,8 | 1,2 |
| carragheen | 1,1 | -0,4 | pinus maritima | 8,7 | 0,2 |
| carum carvi | 3,0 | -0,2 | piper methysticum | 0,6 | -0,4 |
| chamomilla | 6,0 | 0,0 | plantago lanceolata | 0,9 | -0,4 |
| china officinalis | 3,8 | -0,2 | raphanus sativus niger | 0,8 | -0,4 |
| chrysanthellum amer | 2,5 | -0,3 | rhamnus frangula | 4,0 | -0,2 |
| chrysanthemum part | 5,4 | -0,1 | rhamnus purshiana | 1,5 | -0,4 |
| crataegus oxyacantha | 13,2 | 0,5 | rheum officinale | 1,4 | -0,4 |
| cupressus semp | 3,3 | -0,2 | ribes nigrum folia | 3,3 | -0,2 |
| curcuma | 8,8 | 0,2 | ribes nigrum fructus | 1,0 | -0,4 |
| cynara scolymus | 0,0 | -0,5 | rosa canina | 0,2 | -0,5 |
| echinacea angustifolia | 18,6 | 0,9 | rosmarinus officinalis | 1,6 | -0,3 |
| echinacea purpurea | 2,1 | -0,3 | ruscus aculeatus | 2,4 | -0,3 |
| eleutherococcus | 1,9 | -0,3 | salix alba | 2,7 | -0,3 |
| equisetum arvense | 0,8 | -0,4 | salvia officinalis | 0,4 | -0,4 |
| erigeron canadensis | 7,5 | 0,1 | sarsaparilla | 0,7 | -0,4 |
| eschscholtzia cali | 8,4 | 0,2 | senna | 5,4 | -0,1 |
| eucalyptus | 0,9 | -0,4 | solidago virga aurea | 12,7 | 0,5 |
| eugenia caryophyllata | 0,2 | -0,5 | spiraea ulmaria | 6,7 | 0,0 |
| fenugrec | 5,4 | -0,1 | spiruline nebl | 2,2 | -0,3 |
| foeniculum vulgare | 13,2 | 0,5 | spiruline poud | 0,9 | -0,4 |
| fraxinus excelior | 1,4 | -0,4 | taraxacum | 1,2 | -0,4 |
| fucus vesiculosus | 0,2 | -0,5 | thymus vulgaris | 2,5 | -0,3 |
| fumaria officinalis | 4,6 | -0,1 | tilia alburnum | 3,2 | -0,2 |
| gentiana lutea | 2,6 | -0,3 | tilia cordata | 2,8 | -0,3 |
| ginkgo biloba | 1,5 | -0,4 | urtica dioica | 0,5 | -0,4 |
| ginseng | 4,1 | -0,2 | uva ursi | 1,1 | -0,4 |
| glycyrrhiza glabra | 9,4 | 0,2 | vaccinium myrtillus | 1,0 | -0,4 |
| hamamelis virginia | 1,3 | -0,4 | valeriana officinalis | 2,1 | -0,3 |
| harpagophytum proc | 0,6 | -0,4 | vinca minor | 7,1 | 0,1 |
| hibiscus sabdariffa | 2,7 | -0,3 | viola tricolor | 1,7 | -0,3 |
| hieracium pilosella | 16,5 | 0,7 | viscum album | 1,2 | -0,4 |
| humulus lupulus | 63,9 | 4,2 | vitis vinifera | | 8,1 |
| hydrocotyle asiatica | 5,9 | 0,0 | zingiber officinalis | 23,6 | 1,3 |

# Copper

| | | | | |
|---|---|---|---|---|
| vitis vinifera | | 8,1 | salix alba | 2,7 | -0,3 |
| humulus lupulus | 63,9 | 4,2 | tilia cordata | 2,8 | -0,3 |
| orthosiphon stamineus | 27,2 | 1,5 | gentiana lutea | 2,6 | -0,3 |
| zingiber officinalis | 23,6 | 1,3 | marrubium vulgare | 2,6 | -0,3 |
| phaseolus vulgaris | 22,8 | 1,2 | melissa officinalis | 2,6 | -0,3 |
| echinacea angustifolia | 18,6 | 0,9 | chrysanthellum amer | 2,5 | -0,3 |
| hieracium pilosella | 16,5 | 0,7 | ruscus aculeatus | 2,4 | -0,3 |
| aesculus hippocastanum | 16,1 | 0,7 | thymus vulgaris | 2,5 | -0,3 |
| lotus corniculatus | 14,8 | 0,6 | spiruline nebl | 2,2 | -0,3 |
| crataegus oxyacantha | 13,2 | 0,5 | echinacea purpurea | 2,1 | -0,3 |
| foeniculum vulgare | 13,2 | 0,5 | valeriana officinalis | 2,1 | -0,3 |
| solidago virga aurea | 12,7 | 0,5 | carduus marianus | 1,9 | -0,3 |
| hyoscyamus niger | 12,3 | 0,4 | eleutherococcus | 1,9 | -0,3 |
| melilotus officinalis | 10,6 | 0,3 | rosmarinus officinalis | 1,6 | -0,3 |
| glycyrrhiza glabra | 9,4 | 0,2 | viola tricolor | 1,7 | -0,3 |
| curcuma | 8,8 | 0,2 | ginkgo biloba | 1,5 | -0,4 |
| pinus maritima | 8,7 | 0,2 | juniperus communis | 1,5 | -0,4 |
| eschscholtzia cali | 8,4 | 0,2 | rhamnus purshiana | 1,5 | -0,4 |
| erigeron canadensis | 7,5 | 0,1 | fraxinus excelior | 1,4 | -0,4 |
| vinca minor | 7,1 | 0,1 | hamamelis virginia | 1,3 | -0,4 |
| spiraea ulmaria | 6,7 | 0,0 | rheum officinale | 1,4 | -0,4 |
| alchemilla vulgaris | 6,3 | 0,0 | angelica archangelica | 1,2 | -0,4 |
| berberis vulgaris | 6,2 | 0,0 | betula alba | 1,2 | -0,4 |
| chamomilla | 6,0 | 0,0 | nasturtium officinalis | 1,2 | -0,4 |
| hydrocotyle asiatica | 5,9 | 0,0 | taraxacum | 1,2 | -0,4 |
| allium sativum | 5,7 | 0,0 | viscum album | 1,2 | -0,4 |
| caroube | 5,6 | -0,1 | millefolium | 1,1 | -0,4 |
| beta vulgaris | 5,3 | -0,1 | uva ursi | 1,1 | -0,4 |
| senna | 5,4 | -0,1 | carragheen | 1,1 | -0,4 |
| chrysanthemum part | 5,4 | -0,1 | carica papaya | 1,1 | -0,4 |
| fenugrec | 5,4 | -0,1 | eucalyptus | 0,9 | -0,4 |
| ballota foetida | 4,7 | -0,1 | medicago sativa | 0,9 | -0,4 |
| fumaria officinalis | 4,6 | -0,1 | plantago lanceolata | 0,9 | -0,4 |
| ananassa comosus | 4,5 | -0,1 | ribes nigrum fructus | 1,0 | -0,4 |
| arctium lappa | 4,4 | -0,1 | spiruline poud | 0,9 | -0,4 |
| ginseng | 4,1 | -0,2 | vaccinium myrtillus | 1,0 | -0,4 |
| rhamnus frangula | 4,0 | -0,2 | equisetum arvense | 0,8 | -0,4 |
| china officinalis | 3,8 | -0,2 | raphanus sativus niger | 0,8 | -0,4 |
| lespedeza capitata | 3,7 | -0,2 | sarsaparilla | 0,7 | -0,4 |
| lithospermum | 3,7 | -0,2 | boldo fragrans | 0,5 | -0,4 |
| olea europaea | 3,5 | -0,2 | harpagophytum proc | 0,6 | -0,4 |
| acorus calamus | 3,4 | -0,2 | mentha piperita | 0,6 | -0,4 |
| cupressus semp | 3,3 | -0,2 | piper methysticum | 0,6 | -0,4 |
| ribes nigrum folia | 3,3 | -0,2 | urtica dioica | 0,5 | -0,4 |
| hypericum perforatum | 3,1 | -0,2 | salvia officinalis | 0,4 | -0,4 |
| tilia alburnum | 3,2 | -0,2 | eugenia caryophyllata | 0,2 | -0,5 |
| carum carvi | 3,0 | -0,2 | fucus vesiculosus | 0,2 | -0,5 |
| agnus castus | 3,0 | -0,2 | rosa canina | 0,2 | -0,5 |
| hibiscus sabdariffa | 2,7 | -0,3 | althea officinalis | 0,0 | -0,5 |
| passiflora incarnata | 2,7 | -0,3 | cynara scolymus | 0,0 | -0,5 |

# Germanium

| | | | | | | |
|---|---|---|---|---|---|---|
| acorus calamus | 0,0 | -0,5 | hyoscyamus niger | 0,6 | 0,3 |
| aesculus hippocastanum | 0,0 | -0,5 | hypericum perforatum | 1,5 | 1,5 |
| agnus castus | 0,0 | -0,5 | juniperus communis | 0,0 | -0,5 |
| alchemilla vulgaris | 1,0 | 0,8 | lespedeza capitata | 0,0 | -0,5 |
| allium sativum | 0,0 | -0,5 | lithospermum | 0,0 | -0,5 |
| althea officinalis | 0,4 | 0,0 | lotus corniculatus | 0,0 | -0,5 |
| ananassa comosus | 0,0 | -0,5 | marrubium vulgare | 0,0 | -0,5 |
| angelica archangelica | 0,3 | -0,1 | medicago sativa | 0,0 | -0,5 |
| arctium lappa | 0,0 | -0,5 | melilotus officinalis | 0,0 | -0,5 |
| ballota foetida | 0,0 | -0,5 | melissa officinalis | 0,0 | -0,5 |
| berberis vulgaris | 1,4 | 1,3 | mentha piperita | 0,3 | -0,1 |
| beta vulgaris | 0,0 | -0,5 | millefolium | 0,0 | -0,5 |
| betula alba | 0,0 | -0,5 | nasturtium officinalis | 0,0 | -0,5 |
| boldo fragrans | 1,2 | 1,1 | olea europaea | 0,2 | -0,2 |
| carduus marianus | 0,5 | 0,2 | orthosiphon stamineus | 0,0 | -0,5 |
| carica papaya | 0,0 | -0,5 | passiflora incarnata | 0,0 | -0,5 |
| caroube | 0,0 | -0,5 | phaseolus vulgaris | 0,0 | -0,5 |
| carragheen | 4,4 | 5,2 | pinus maritima | 0,0 | -0,5 |
| carum carvi | 1,0 | 0,8 | piper methysticum | 2,0 | 2,1 |
| chamomilla | 0,1 | -0,4 | plantago lanceolata | 0,1 | -0,3 |
| china officinalis | 0,0 | -0,5 | raphanus sativus niger | 1,6 | 1,6 |
| chrysanthellum amer | 0,7 | 0,4 | rhamnus frangula | 0,0 | -0,5 |
| chrysanthemum part | 0,7 | 0,4 | rhamnus purshiana | 2,4 | 2,6 |
| crataegus oxyacantha | 0,0 | -0,5 | rheum officinale | 0,3 | -0,1 |
| cupressus semp | 0,0 | -0,5 | ribes nigrum folia | 0,0 | -0,5 |
| curcuma | 0,0 | -0,5 | ribes nigrum fructus | 0,0 | -0,5 |
| cynara scolymus | 0,0 | -0,5 | rosa canina | 1,7 | 1,7 |
| echinacea angustifolia | 0,0 | -0,5 | rosmarinus officinalis | 0,0 | -0,5 |
| echinacea purpurea | 0,0 | -0,5 | ruscus aculeatus | 3,3 | 3,8 |
| eleutherococcus | 0,3 | -0,1 | salix alba | 0,0 | -0,5 |
| equisetum arvense | 0,0 | -0,5 | salvia officinalis | 0,7 | 0,4 |
| erigeron canadensis | 0,0 | -0,5 | sarsaparilla | 0,0 | -0,5 |
| eschscholtzia cali | 0,0 | -0,5 | senna | 0,0 | -0,5 |
| eucalyptus | 0,0 | -0,5 | solidago virga aurea | 0,0 | -0,5 |
| eugenia caryophyllata | 0,0 | -0,5 | spiraea ulmaria | 0,0 | -0,5 |
| fenugrec | 0,0 | -0,5 | spiruline nebl | 0,0 | -0,5 |
| foeniculum vulgare | 0,0 | -0,5 | spiruline poud | 1,1 | 1,0 |
| fraxinus excelior | 0,0 | -0,5 | taraxacum | 1,7 | 1,7 |
| fucus vesiculosus | 0,0 | -0,5 | thymus vulgaris | 0,0 | -0,5 |
| fumaria officinalis | 0,0 | -0,5 | tilia alburnum | 0,0 | -0,5 |
| gentiana lutea | 2,6 | 2,9 | tilia cordata | 0,0 | -0,5 |
| ginkgo biloba | 0,0 | -0,5 | urtica dioica | 0,0 | -0,5 |
| ginseng | 1,6 | 1,6 | uva ursi | 0,3 | -0,1 |
| glycyrrhiza glabra | 1,6 | 1,6 | vaccinium myrtillus | 0,4 | 0,0 |
| hamamelis virginia | 0,6 | 0,3 | valeriana officinalis | 0,0 | -0,5 |
| harpagophytum proc | 0,4 | 0,0 | vinca minor | 0,0 | -0,5 |
| hibiscus sabdariffa | 0,0 | -0,5 | viola tricolor | 0,0 | -0,5 |
| hieracium pilosella | 0,0 | -0,5 | viscum album | 0,0 | -0,5 |
| humulus lupulus | 0,0 | -0,5 | vitis vinifera | 0,0 | -0,5 |
| hydrocotyle asiatica | 0,0 | -0,5 | zingiber officinalis | 0,0 | -0,5 |

# Germanium

| | | | | |
|---|---|---|---|---|
| carragheen | 4,4 | 5,2 | echinacea purpurea | 0,0 | -0,5 |
| ruscus aculeatus | 3,3 | 3,8 | equisetum arvense | 0,0 | -0,5 |
| gentiana lutea | 2,6 | 2,9 | erigeron canadensis | 0,0 | -0,5 |
| rhamnus purshiana | 2,4 | 2,6 | eschscholtzia cali | 0,0 | -0,5 |
| piper methysticum | 2,0 | 2,1 | eucalyptus | 0,0 | -0,5 |
| rosa canina | 1,7 | 1,7 | eugenia caryophyllata | 0,0 | -0,5 |
| taraxacum | 1,7 | 1,7 | fenugrec | 0,0 | -0,5 |
| glycyrrhiza glabra | 1,6 | 1,6 | foeniculum vulgare | 0,0 | -0,5 |
| ginseng | 1,6 | 1,6 | fraxinus excelior | 0,0 | -0,5 |
| raphanus sativus niger | 1,6 | 1,6 | fucus vesiculosus | 0,0 | -0,5 |
| hypericum perforatum | 1,5 | 1,5 | fumaria officinalis | 0,0 | -0,5 |
| berberis vulgaris | 1,4 | 1,3 | ginkgo biloba | 0,0 | -0,5 |
| boldo fragrans | 1,2 | 1,1 | hibiscus sabdariffa | 0,0 | -0,5 |
| spiruline poud | 1,1 | 1,0 | hieracium pilosella | 0,0 | -0,5 |
| alchemilla vulgaris | 1,0 | 0,8 | humulus lupulus | 0,0 | -0,5 |
| carum carvi | 1,0 | 0,8 | juniperus communis | 0,0 | -0,5 |
| chrysanthellum amer | 0,7 | 0,4 | lespedeza capitata | 0,0 | -0,5 |
| chrysanthemum part | 0,7 | 0,4 | lithospermum | 0,0 | -0,5 |
| salvia officinalis | 0,7 | 0,4 | lotus corniculatus | 0,0 | -0,5 |
| hamamelis virginia | 0,6 | 0,3 | marrubium vulgare | 0,0 | -0,5 |
| hyoscyamus niger | 0,6 | 0,3 | medicago sativa | 0,0 | -0,5 |
| carduus marianus | 0,5 | 0,2 | melilotus officinalis | 0,0 | -0,5 |
| althea officinalis | 0,4 | 0,0 | carica papaya | 0,0 | -0,5 |
| harpagophytum proc | 0,4 | 0,0 | melissa officinalis | 0,0 | -0,5 |
| vaccinium myrtillus | 0,4 | 0,0 | nasturtium officinalis | 0,0 | -0,5 |
| angelica archangelica | 0,3 | -0,1 | orthosiphon stamineus | 0,0 | -0,5 |
| uva ursi | 0,3 | -0,1 | passiflora incarnata | 0,0 | -0,5 |
| eleutherococcus | 0,3 | -0,1 | phaseolus vulgaris | 0,0 | -0,5 |
| mentha piperita | 0,3 | -0,1 | pinus maritima | 0,0 | -0,5 |
| rheum officinale | 0,3 | -0,1 | china officinalis | 0,0 | -0,5 |
| olea europaea | 0,2 | -0,2 | rhamnus frangula | 0,0 | -0,5 |
| plantago lanceolata | 0,1 | -0,3 | ribes nigrum folia | 0,0 | -0,5 |
| chamomilla | 0,1 | -0,4 | ribes nigrum fructus | 0,0 | -0,5 |
| millefolium | 0,0 | -0,5 | rosmarinus officinalis | 0,0 | -0,5 |
| acorus calamus | 0,0 | -0,5 | salix alba | 0,0 | -0,5 |
| aesculus hippocastanum | 0,0 | -0,5 | sarsaparilla | 0,0 | -0,5 |
| allium sativum | 0,0 | -0,5 | solidago virga aurea | 0,0 | -0,5 |
| ananassa comosus | 0,0 | -0,5 | spiraea ulmaria | 0,0 | -0,5 |
| arctium lappa | 0,0 | -0,5 | spiruline nebl | 0,0 | -0,5 |
| ballota foetida | 0,0 | -0,5 | thymus vulgaris | 0,0 | -0,5 |
| beta vulgaris | 0,0 | -0,5 | tilia alburnum | 0,0 | -0,5 |
| betula alba | 0,0 | -0,5 | tilia cordata | 0,0 | -0,5 |
| caroube | 0,0 | -0,5 | urtica dioica | 0,0 | -0,5 |
| senna | 0,0 | -0,5 | valeriana officinalis | 0,0 | -0,5 |
| hydrocotyle asiatica | 0,0 | -0,5 | vinca minor | 0,0 | -0,5 |
| crataegus oxyacantha | 0,0 | -0,5 | viola tricolor | 0,0 | -0,5 |
| cupressus semp | 0,0 | -0,5 | viscum album | 0,0 | -0,5 |
| curcuma | 0,0 | -0,5 | agnus castus | 0,0 | -0,5 |
| cynara scolymus | 0,0 | -0,5 | vitis vinifera | 0,0 | -0,5 |
| echinacea angustifolia | 0,0 | -0,5 | zingiber officinalis | 0,0 | -0,5 |

# Iron

| | | | | | |
|---|---|---|---|---|---|
| acorus calamus | 230,0 | 1,1 | hyoscyamus niger | 27,8 | -0,3 |
| aesculus hippocastanum | 2,6 | -0,5 | hypericum perforatum | 3,1 | -0,5 |
| agnus castus | 36,4 | -0,2 | juniperus communis | 22,1 | -0,3 |
| alchemilla vulgaris | 15,4 | -0,4 | lespedeza capitata | 14,1 | -0,4 |
| allium sativum | 14,5 | -0,4 | lithospermum | 88,1 | 0,1 |
| althea officinalis | 20,6 | -0,3 | lotus corniculatus | 96,5 | 0,2 |
| ananassa comosus | 12,2 | -0,4 | marrubium vulgare | 52,0 | -0,1 |
| angelica archangelica | 27,1 | -0,3 | medicago sativa | 27,6 | -0,3 |
| arctium lappa | 71,1 | 0,0 | melilotus officinalis | 12,2 | -0,4 |
| ballota foetida | 82,0 | 0,1 | melissa officinalis | 4,0 | -0,5 |
| berberis vulgaris | 17,8 | -0,4 | mentha piperita | 24,2 | -0,3 |
| beta vulgaris | 59,3 | -0,1 | millefolium | 17,1 | -0,4 |
| betula alba | 21,8 | -0,3 | nasturtium officinalis | 293,2 | 1,5 |
| boldo fragrans | 14,5 | -0,4 | olea europaea | 13,3 | -0,4 |
| carduus marianus | 21,5 | -0,3 | orthosiphon stamineus | 14,3 | -0,4 |
| carica papaya | 10,9 | -0,4 | passiflora incarnata | 27,1 | -0,3 |
| caroube | 1178,0 | 7,6 | phaseolus vulgaris | 33,4 | -0,2 |
| carragheen | 168,6 | 0,7 | pinus maritima | 41,0 | -0,2 |
| carum carvi | 35,5 | -0,2 | piper methysticum | 1,1 | -0,5 |
| chamomilla | 20,7 | -0,3 | plantago lanceolata | 37,3 | -0,2 |
| china officinalis | 46,5 | -0,2 | raphanus sativus niger | 32,7 | -0,3 |
| chrysanthellum amer | 110,0 | 0,3 | rhamnus frangula | 12,1 | -0,4 |
| chrysanthemum part | 2,3 | -0,5 | rhamnus purshiana | 20,2 | -0,3 |
| crataegus oxyacantha | 8,3 | -0,4 | rheum officinale | 5,7 | -0,4 |
| cupressus semp | 146,5 | 0,5 | ribes nigrum folia | 30,0 | -0,3 |
| curcuma | 58,9 | -0,1 | ribes nigrum fructus | 28,5 | -0,3 |
| cynara scolymus | 23,7 | -0,3 | rosa canina | 4,2 | -0,4 |
| echinacea angustifolia | 8,5 | -0,4 | rosmarinus officinalis | 61,7 | -0,1 |
| echinacea purpurea | 11,7 | -0,4 | ruscus aculeatus | 9,3 | -0,4 |
| eleutherococcus | 4,5 | -0,4 | salix alba | 134,0 | 0,4 |
| equisetum arvense | 13,7 | -0,4 | salvia officinalis | 29,2 | -0,3 |
| erigeron canadensis | 4,7 | -0,4 | sarsaparilla | 26,7 | -0,3 |
| eschscholtzia cali | 285,2 | 1,5 | senna | 9,8 | -0,4 |
| eucalyptus | 1,8 | -0,5 | solidago virga aurea | 12,9 | -0,4 |
| eugenia caryophyllata | 11,0 | -0,4 | spiraea ulmaria | 6,9 | -0,4 |
| fenugrec | 18,9 | -0,3 | spiruline nebl | 30,2 | -0,3 |
| foeniculum vulgare | 31,1 | -0,3 | spiruline poud | 315,5 | 1,7 |
| fraxinus excelior | 205,8 | 0,9 | taraxacum | 45,6 | -0,2 |
| fucus vesiculosus | 8,7 | -0,4 | thymus vulgaris | 45,8 | -0,2 |
| fumaria officinalis | 83,4 | 0,1 | tilia alburnum | 35,9 | -0,2 |
| gentiana lutea | 74,4 | 0,0 | tilia cordata | 4,4 | -0,4 |
| ginkgo biloba | 37,7 | -0,2 | urtica dioica | 32,5 | -0,3 |
| ginseng | 2,9 | -0,5 | uva ursi | 21,5 | -0,3 |
| glycyrrhiza glabra | 181,6 | 0,8 | vaccinium myrtillus | 19,0 | -0,3 |
| hamamelis virginia | 12,1 | -0,4 | valeriana officinalis | 57,3 | -0,1 |
| harpagophytum proc | 8,0 | -0,4 | vinca minor | 29,7 | -0,3 |
| hibiscus sabdariffa | 95,3 | 0,2 | viola tricolor | 305,1 | 1,6 |
| hieracium pilosella | 651,9 | 4,0 | viscum album | 88,8 | 0,1 |
| humulus lupulus | 5,4 | -0,4 | vitis vinifera | 277,6 | 1,4 |
| hydrocotyle asiatica | 33,0 | -0,3 | zingiber officinalis | 78,6 | 0,1 |

# Iron

| | | | | | |
|---|---|---|---|---|---|
| caroube | 1178,0 | 7,6 | passiflora incarnata | 27,1 | -0,3 |
| hieracium pilosella | 651,9 | 4,0 | sarsaparilla | 26,7 | -0,3 |
| spiruline poud | 315,5 | 1,7 | cynara scolymus | 23,7 | -0,3 |
| viola tricolor | 305,1 | 1,6 | mentha piperita | 24,2 | -0,3 |
| nasturtium officinalis | 293,2 | 1,5 | betula alba | 21,8 | -0,3 |
| eschscholtzia cali | 285,2 | 1,5 | juniperus communis | 22,1 | -0,3 |
| vitis vinifera | 277,6 | 1,4 | althea officinalis | 20,6 | -0,3 |
| acorus calamus | 230,0 | 1,1 | uva ursi | 21,5 | -0,3 |
| fraxinus excelior | 205,8 | 0,9 | carduus marianus | 21,5 | -0,3 |
| glycyrrhiza glabra | 181,6 | 0,8 | chamomilla | 20,7 | -0,3 |
| carragheen | 168,6 | 0,7 | rhamnus purshiana | 20,2 | -0,3 |
| cupressus semp | 146,5 | 0,5 | fenugrec | 18,9 | -0,3 |
| salix alba | 134,0 | 0,4 | vaccinium myrtillus | 19,0 | -0,3 |
| chrysanthellum amer | 110,0 | 0,3 | berberis vulgaris | 17,8 | -0,4 |
| lotus corniculatus | 96,5 | 0,2 | millefolium | 17,1 | -0,4 |
| hibiscus sabdariffa | 95,3 | 0,2 | alchemilla vulgaris | 15,4 | -0,4 |
| lithospermum | 88,1 | 0,1 | allium sativum | 14,5 | -0,4 |
| viscum album | 88,8 | 0,1 | boldo fragrans | 14,5 | -0,4 |
| fumaria officinalis | 83,4 | 0,1 | olea europaea | 13,3 | -0,4 |
| ballota foetida | 82,0 | 0,1 | equisetum arvense | 13,7 | -0,4 |
| zingiber officinalis | 78,6 | 0,1 | lespedeza capitata | 14,1 | -0,4 |
| gentiana lutea | 74,4 | 0,0 | orthosiphon stamineus | 14,3 | -0,4 |
| arctium lappa | 71,1 | 0,0 | solidago virga aurea | 12,9 | -0,4 |
| rosmarinus officinalis | 61,7 | -0,1 | ananassa comosus | 12,2 | -0,4 |
| beta vulgaris | 59,3 | -0,1 | echinacea purpurea | 11,7 | -0,4 |
| curcuma | 58,9 | -0,1 | hamamelis virginia | 12,1 | -0,4 |
| valeriana officinalis | 57,3 | -0,1 | melilotus officinalis | 12,2 | -0,4 |
| marrubium vulgare | 52,0 | -0,1 | rhamnus frangula | 12,1 | -0,4 |
| china officinalis | 46,5 | -0,2 | eugenia caryophyllata | 11,0 | -0,4 |
| taraxacum | 45,6 | -0,2 | carica papaya | 10,9 | -0,4 |
| thymus vulgaris | 45,8 | -0,2 | senna | 9,8 | -0,4 |
| pinus maritima | 41,0 | -0,2 | fucus vesiculosus | 8,7 | -0,4 |
| ginkgo biloba | 37,7 | -0,2 | ruscus aculeatus | 9,3 | -0,4 |
| plantago lanceolata | 37,3 | -0,2 | crataegus oxyacantha | 8,3 | -0,4 |
| agnus castus | 36,4 | -0,2 | echinacea angustifolia | 8,5 | -0,4 |
| carum carvi | 35,5 | -0,2 | harpagophytum proc | 8,0 | -0,4 |
| tilia alburnum | 35,9 | -0,2 | rheum officinale | 5,7 | -0,4 |
| phaseolus vulgaris | 33,4 | -0,2 | spiraea ulmaria | 6,9 | -0,4 |
| hydrocotyle asiatica | 33,0 | -0,3 | eleutherococcus | 4,5 | -0,4 |
| raphanus sativus niger | 32,7 | -0,3 | erigeron canadensis | 4,7 | -0,4 |
| urtica dioica | 32,5 | -0,3 | humulus lupulus | 5,4 | -0,4 |
| foeniculum vulgare | 31,1 | -0,3 | rosa canina | 4,2 | -0,4 |
| ribes nigrum folia | 30,0 | -0,3 | tilia cordata | 4,4 | -0,4 |
| salvia officinalis | 29,2 | -0,3 | hypericum perforatum | 3,1 | -0,5 |
| spiruline nebl | 30,2 | -0,3 | melissa officinalis | 4,0 | -0,5 |
| vinca minor | 29,7 | -0,3 | ginseng | 2,9 | -0,5 |
| hyoscyamus niger | 27,8 | -0,3 | aesculus hippocastanum | 2,6 | -0,5 |
| medicago sativa | 27,6 | -0,3 | chrysanthemum part | 2,3 | -0,5 |
| ribes nigrum fructus | 28,5 | -0,3 | eucalyptus | 1,8 | -0,5 |
| angelica archangelica | 27,1 | -0,3 | piper methysticum | 1,1 | -0,5 |

# Lead

| | | | |
|---|---|---|---|
| acorus calamus | 0,0 | -0,4 | |
| aesculus hippocastanum | 0,0 | -0,4 | |
| agnus castus | 0,0 | -0,4 | |
| alchemilla vulgaris | 0,0 | -0,4 | |
| allium sativum | 0,0 | -0,4 | |
| althea officinalis | 0,0 | -0,4 | |
| ananassa comosus | 0,0 | -0,4 | |
| angelica archangelica | 0,8 | 0,4 | |
| arctium lappa | 0,7 | 0,3 | |
| ballota foetida | 0,5 | 0,1 | |
| berberis vulgaris | 0,0 | -0,4 | |
| beta vulgaris | 0,0 | -0,4 | |
| betula alba | 1,0 | 0,6 | |
| boldo fragrans | 0,0 | -0,4 | |
| carduus marianus | 0,0 | -0,4 | |
| carica papaya | 4,5 | 4,1 | |
| caroube | 0,0 | -0,4 | |
| carragheen | 0,0 | -0,4 | |
| carum carvi | 0,0 | -0,4 | |
| chamomilla | 0,0 | -0,4 | |
| china officinalis | 2,7 | 2,3 | |
| chrysanthellum amer | 0,0 | -0,4 | |
| chrysanthemum part | 0,0 | -0,4 | |
| crataegus oxyacantha | 0,0 | -0,4 | |
| cupressus semp | 0,0 | -0,4 | |
| curcuma | 0,0 | -0,4 | |
| cynara scolymus | 0,0 | -0,4 | |
| echinacea angustifolia | 0,0 | -0,4 | |
| echinacea purpurea | 0,0 | -0,4 | |
| eleutherococcus | 0,0 | -0,4 | |
| equisetum arvense | 0,0 | -0,4 | |
| erigeron canadensis | 0,0 | -0,4 | |
| eschscholtzia cali | 0,0 | -0,4 | |
| eucalyptus | 0,9 | 0,5 | |
| eugenia caryophyllata | 0,0 | -0,4 | |
| fenugrec | 0,0 | -0,4 | |
| foeniculum vulgare | 0,3 | -0,1 | |
| fraxinus excelior | 3,2 | 2,8 | |
| fucus vesiculosus | 0,0 | -0,4 | |
| fumaria officinalis | 0,0 | -0,4 | |
| gentiana lutea | 0,0 | -0,4 | |
| ginkgo biloba | 0,0 | -0,4 | |
| ginseng | 0,0 | -0,4 | |
| glycyrrhiza glabra | 0,0 | -0,4 | |
| hamamelis virginia | 1,5 | 1,1 | |
| harpagophytum proc | 0,0 | -0,4 | |
| hibiscus sabdariffa | 0,0 | -0,4 | |
| hieracium pilosella | 1,9 | 1,5 | |
| humulus lupulus | 0,0 | -0,4 | |
| hydrocotyle asiatica | 0,6 | 0,2 | |

| | | |
|---|---|---|
| hyoscyamus niger | 0,0 | -0,4 |
| hypericum perforatum | 0,0 | -0,4 |
| juniperus communis | 0,0 | -0,4 |
| lespedeza capitata | 0,0 | -0,4 |
| lithospermum | 4,4 | 4,0 |
| lotus corniculatus | 0,1 | -0,4 |
| marrubium vulgare | 1,1 | 0,7 |
| medicago sativa | 0,0 | -0,4 |
| melilotus officinalis | 2,8 | 2,4 |
| melissa officinalis | 0,0 | -0,4 |
| mentha piperita | 0,0 | -0,4 |
| millefolium | 0,0 | -0,4 |
| nasturtium officinalis | 0,0 | -0,4 |
| olea europaea | 0,0 | -0,4 |
| orthosiphon stamineus | 0,0 | -0,4 |
| passiflora incarnata | 1,7 | 1,3 |
| phaseolus vulgaris | 3,5 | 3,1 |
| pinus maritima | 0,0 | -0,4 |
| piper methysticum | 0,0 | -0,4 |
| plantago lanceolata | 0,0 | -0,4 |
| raphanus sativus niger | 0,0 | -0,4 |
| rhamnus frangula | 0,0 | -0,4 |
| rhamnus purshiana | 0,0 | -0,4 |
| rheum officinale | 0,0 | -0,4 |
| ribes nigrum folia | 0,0 | -0,4 |
| ribes nigrum fructus | 0,0 | -0,4 |
| rosa canina | 2,8 | 2,4 |
| rosmarinus officinalis | 0,1 | -0,3 |
| ruscus aculeatus | 0,0 | -0,4 |
| salix alba | 0,0 | -0,4 |
| salvia officinalis | 0,0 | -0,4 |
| sarsaparilla | 0,0 | -0,4 |
| senna | 0,0 | -0,4 |
| solidago virga aurea | 2,5 | 2,1 |
| spiraea ulmaria | 0,0 | -0,4 |
| spiruline nebl | 0,0 | -0,4 |
| spiruline poud | 0,0 | -0,4 |
| taraxacum | 3,3 | 2,9 |
| thymus vulgaris | 0,0 | -0,4 |
| tilia alburnum | 0,0 | -0,4 |
| tilia cordata | 0,0 | -0,4 |
| urtica dioica | 0,5 | 0,1 |
| uva ursi | 0,0 | -0,4 |
| vaccinium myrtillus | 1,2 | 0,8 |
| valeriana officinalis | 0,0 | -0,4 |
| vinca minor | 0,0 | -0,4 |
| viola tricolor | 1,2 | 0,8 |
| viscum album | 0,0 | -0,4 |
| vitis vinifera | 0,0 | -0,4 |
| zingiber officinalis | 0,4 | 0,0 |

# Lead

| | | | | |
|---|---|---|---|---|
| carica papaya | 4,5 | 4,1 | eleutherococcus | 0,0 | -0,4 |
| lithospermum | 4,4 | 4,0 | equisetum arvense | 0,0 | -0,4 |
| phaseolus vulgaris | 3,5 | 3,1 | erigeron canadensis | 0,0 | -0,4 |
| taraxacum | 3,3 | 2,9 | eschscholtzia cali | 0,0 | -0,4 |
| fraxinus excelior | 3,2 | 2,8 | eugenia caryophyllata | 0,0 | -0,4 |
| melilotus officinalis | 2,8 | 2,4 | fenugrec | 0,0 | -0,4 |
| rosa canina | 2,8 | 2,4 | fucus vesiculosus | 0,0 | -0,4 |
| china officinalis | 2,7 | 2,3 | fumaria officinalis | 0,0 | -0,4 |
| solidago virga aurea | 2,5 | 2,1 | gentiana lutea | 0,0 | -0,4 |
| hieracium pilosella | 1,9 | 1,5 | ginkgo biloba | 0,0 | -0,4 |
| passiflora incarnata | 1,7 | 1,3 | glycyrrhiza glabra | 0,0 | -0,4 |
| hamamelis virginia | 1,5 | 1,1 | boldo fragrans | 0,0 | -0,4 |
| vaccinium myrtillus | 1,2 | 0,8 | hibiscus sabdariffa | 0,0 | -0,4 |
| viola tricolor | 1,2 | 0,8 | harpagophytum proc | 0,0 | -0,4 |
| marrubium vulgare | 1,1 | 0,7 | humulus lupulus | 0,0 | -0,4 |
| betula alba | 1,0 | 0,6 | hyoscyamus niger | 0,0 | -0,4 |
| eucalyptus | 0,9 | 0,5 | hypericum perforatum | 0,0 | -0,4 |
| angelica archangelica | 0,8 | 0,4 | juniperus communis | 0,0 | -0,4 |
| arctium lappa | 0,7 | 0,3 | lespedeza capitata | 0,0 | -0,4 |
| hydrocotyle asiatica | 0,6 | 0,2 | chamomilla | 0,0 | -0,4 |
| ballota foetida | 0,5 | 0,1 | medicago sativa | 0,0 | -0,4 |
| urtica dioica | 0,5 | 0,1 | melissa officinalis | 0,0 | -0,4 |
| zingiber officinalis | 0,4 | 0,0 | mentha piperita | 0,0 | -0,4 |
| foeniculum vulgare | 0,3 | -0,1 | nasturtium officinalis | 0,0 | -0,4 |
| rosmarinus officinalis | 0,1 | -0,3 | orthosiphon stamineus | 0,0 | -0,4 |
| lotus corniculatus | 0,1 | -0,4 | ginseng | 0,0 | -0,4 |
| millefolium | 0,0 | -0,4 | pinus maritima | 0,0 | -0,4 |
| acorus calamus | 0,0 | -0,4 | piper methysticum | 0,0 | -0,4 |
| aesculus hippocastanum | 0,0 | -0,4 | plantago lanceolata | 0,0 | -0,4 |
| alchemilla vulgaris | 0,0 | -0,4 | raphanus sativus niger | 0,0 | -0,4 |
| allium sativum | 0,0 | -0,4 | rhamnus frangula | 0,0 | -0,4 |
| althea officinalis | 0,0 | -0,4 | rhamnus purshiana | 0,0 | -0,4 |
| ananassa comosus | 0,0 | -0,4 | rheum officinale | 0,0 | -0,4 |
| uva ursi | 0,0 | -0,4 | ribes nigrum folia | 0,0 | -0,4 |
| berberis vulgaris | 0,0 | -0,4 | ribes nigrum fructus | 0,0 | -0,4 |
| beta vulgaris | 0,0 | -0,4 | ruscus aculeatus | 0,0 | -0,4 |
| carduus marianus | 0,0 | -0,4 | salix alba | 0,0 | -0,4 |
| caroube | 0,0 | -0,4 | salvia officinalis | 0,0 | -0,4 |
| carragheen | 0,0 | -0,4 | sarsaparilla | 0,0 | -0,4 |
| carum carvi | 0,0 | -0,4 | spiraea ulmaria | 0,0 | -0,4 |
| senna | 0,0 | -0,4 | spiruline nebl | 0,0 | -0,4 |
| chrysanthellum amer | 0,0 | -0,4 | spiruline poud | 0,0 | -0,4 |
| chrysanthemum part | 0,0 | -0,4 | thymus vulgaris | 0,0 | -0,4 |
| crataegus oxyacantha | 0,0 | -0,4 | tilia alburnum | 0,0 | -0,4 |
| cupressus semp | 0,0 | -0,4 | tilia cordata | 0,0 | -0,4 |
| curcuma | 0,0 | -0,4 | valeriana officinalis | 0,0 | -0,4 |
| cynara scolymus | 0,0 | -0,4 | vinca minor | 0,0 | -0,4 |
| echinacea angustifolia | 0,0 | -0,4 | viscum album | 0,0 | -0,4 |
| echinacea purpurea | 0,0 | -0,4 | agnus castus | 0,0 | -0,4 |
| olea europaea | 0,0 | -0,4 | vitis vinifera | 0,0 | -0,4 |

# Lithium

| | | | | | | |
|---|---|---|---|---|---|---|
| acorus calamus | 1,6 | 0,5 | hyoscyamus niger | 11,9 | 6,8 |
| aesculus hippocastanum | 0,0 | -0,5 | hypericum perforatum | 0,0 | -0,5 |
| agnus castus | 0,0 | -0,5 | juniperus communis | 1,4 | 0,4 |
| alchemilla vulgaris | 0,7 | 0,0 | lespedeza capitata | 0,0 | -0,5 |
| allium sativum | 0,4 | -0,2 | lithospermum | 0,5 | -0,2 |
| althea officinalis | 0,4 | -0,2 | lotus corniculatus | 0,5 | -0,2 |
| ananassa comosus | 0,0 | -0,5 | marrubium vulgare | 1,1 | 0,2 |
| angelica archangelica | 0,0 | -0,5 | medicago sativa | 0,9 | 0,1 |
| arctium lappa | 0,5 | -0,2 | melilotus officinalis | 0,4 | -0,2 |
| ballota foetida | 1,0 | 0,2 | melissa officinalis | 0,0 | -0,5 |
| berberis vulgaris | 0,5 | -0,2 | mentha piperita | 1,3 | 0,3 |
| beta vulgaris | 1,1 | 0,2 | millefolium | 2,3 | 1,0 |
| betula alba | 1,1 | 0,2 | nasturtium officinalis | 0,6 | -0,1 |
| boldo fragrans | 0,2 | -0,3 | olea europaea | 1,0 | 0,2 |
| carduus marianus | 0,3 | -0,3 | orthosiphon stamineus | 0,0 | -0,5 |
| carica papaya | 0,3 | -0,3 | passiflora incarnata | 0,4 | -0,2 |
| caroube | 0,0 | -0,5 | phaseolus vulgaris | 0,4 | -0,2 |
| carragheen | 0,0 | -0,5 | pinus maritima | 1,1 | 0,2 |
| carum carvi | 0,0 | -0,5 | piper methysticum | 0,0 | -0,4 |
| chamomilla | 0,9 | 0,1 | plantago lanceolata | 0,0 | -0,5 |
| china officinalis | 0,1 | -0,4 | raphanus sativus niger | 0,0 | -0,5 |
| chrysanthellum amer | 0,4 | -0,2 | rhamnus frangula | 0,7 | 0,0 |
| chrysanthemum part | 0,0 | -0,5 | rhamnus purshiana | 0,2 | -0,3 |
| crataegus oxyacantha | 1,7 | 0,6 | rheum officinale | 0,0 | -0,5 |
| cupressus semp | 1,7 | 0,6 | ribes nigrum folia | 1,3 | 0,3 |
| curcuma | 0,0 | -0,5 | ribes nigrum fructus | 0,8 | 0,0 |
| cynara scolymus | 0,0 | -0,5 | rosa canina | 0,4 | -0,2 |
| echinacea angustifolia | 0,0 | -0,5 | rosmarinus officinalis | 1,3 | 0,3 |
| echinacea purpurea | 0,2 | -0,3 | ruscus aculeatus | 0,0 | -0,5 |
| eleutherococcus | 0,1 | -0,4 | salix alba | 0,0 | -0,5 |
| equisetum arvense | 0,0 | -0,5 | salvia officinalis | 0,7 | 0,0 |
| erigeron canadensis | 0,3 | -0,3 | sarsaparilla | 0,0 | -0,5 |
| eschscholtzia cali | 1,8 | 0,6 | senna | 0,0 | -0,5 |
| eucalyptus | 1,6 | 0,5 | solidago virga aurea | 0,3 | -0,3 |
| eugenia caryophyllata | 0,0 | -0,5 | spiraea ulmaria | 9,7 | 5,5 |
| fenugrec | 0,1 | -0,4 | spiruline nebl | 0,5 | -0,2 |
| foeniculum vulgare | 5,2 | 2,7 | spiruline poud | 0,4 | -0,2 |
| fraxinus excelior | 2,1 | 0,8 | taraxacum | 1,5 | 0,5 |
| fucus vesiculosus | 0,4 | -0,2 | thymus vulgaris | 0,2 | -0,3 |
| fumaria officinalis | 0,0 | -0,5 | tilia alburnum | 0,2 | -0,3 |
| gentiana lutea | 0,0 | -0,5 | tilia cordata | 0,0 | -0,5 |
| ginkgo biloba | 2,7 | 1,2 | urtica dioica | 0,0 | -0,5 |
| ginseng | 0,0 | -0,5 | uva ursi | 0,0 | -0,5 |
| glycyrrhiza glabra | 1,3 | 0,3 | vaccinium myrtillus | 0,5 | -0,2 |
| hamamelis virginia | 0,1 | -0,4 | valeriana officinalis | 0,8 | 0,0 |
| harpagophytum proc | 0,0 | -0,5 | vinca minor | 0,0 | -0,5 |
| hibiscus sabdariffa | 0,3 | -0,3 | viola tricolor | 0,0 | -0,5 |
| hieracium pilosella | 0,3 | -0,3 | viscum album | 0,0 | -0,5 |
| humulus lupulus | 0,4 | -0,2 | vitis vinifera | 1,2 | 0,3 |
| hydrocotyle asiatica | 0,1 | -0,4 | zingiber officinalis | 0,4 | -0,2 |

# Lithium

| | | | | | |
|---|---|---|---|---|---|
| hyoscyamus niger | 11,9 | 6,8 | erigeron canadensis | 0,3 | -0,3 |
| spiraea ulmaria | 9,7 | 5,5 | hibiscus sabdariffa | 0,3 | -0,3 |
| foeniculum vulgare | 5,2 | 2,7 | hieracium pilosella | 0,3 | -0,3 |
| ginkgo biloba | 2,7 | 1,2 | carica papaya | 0,3 | -0,3 |
| millefolium | 2,3 | 1,0 | solidago virga aurea | 0,3 | -0,3 |
| fraxinus excelior | 2,1 | 0,8 | echinacea purpurea | 0,2 | -0,3 |
| eschscholtzia cali | 1,8 | 0,6 | boldo fragrans | 0,2 | -0,3 |
| crataegus oxyacantha | 1,7 | 0,6 | rhamnus purshiana | 0,2 | -0,3 |
| cupressus semp | 1,7 | 0,6 | thymus vulgaris | 0,2 | -0,3 |
| acorus calamus | 1,6 | 0,5 | tilia alburnum | 0,2 | -0,3 |
| eucalyptus | 1,6 | 0,5 | hydrocotyle asiatica | 0,1 | -0,4 |
| taraxacum | 1,5 | 0,5 | eleutherococcus | 0,1 | -0,4 |
| juniperus communis | 1,4 | 0,4 | fenugrec | 0,1 | -0,4 |
| glycyrrhiza glabra | 1,3 | 0,3 | china officinalis | 0,1 | -0,4 |
| mentha piperita | 1,3 | 0,3 | hamamelis virginia | 0,1 | -0,4 |
| ribes nigrum folia | 1,3 | 0,3 | piper methysticum | 0,0 | -0,4 |
| rosmarinus officinalis | 1,3 | 0,3 | aesculus hippocastanum | 0,0 | -0,5 |
| vitis vinifera | 1,2 | 0,3 | ananassa comosus | 0,0 | -0,5 |
| beta vulgaris | 1,1 | 0,2 | angelica archangelica | 0,0 | -0,5 |
| betula alba | 1,1 | 0,2 | uva ursi | 0,0 | -0,5 |
| marrubium vulgare | 1,1 | 0,2 | caroube | 0,0 | -0,5 |
| pinus maritima | 1,1 | 0,2 | carragheen | 0,0 | -0,5 |
| ballota foetida | 1,0 | 0,2 | carum carvi | 0,0 | -0,5 |
| olea europaea | 1,0 | 0,2 | senna | 0,0 | -0,5 |
| chamomilla | 0,9 | 0,1 | chrysanthemum part | 0,0 | -0,5 |
| medicago sativa | 0,9 | 0,1 | curcuma | 0,0 | -0,5 |
| ribes nigrum fructus | 0,8 | 0,0 | cynara scolymus | 0,0 | -0,5 |
| valeriana officinalis | 0,8 | 0,0 | echinacea angustifolia | 0,0 | -0,5 |
| alchemilla vulgaris | 0,7 | 0,0 | equisetum arvense | 0,0 | -0,5 |
| rhamnus frangula | 0,7 | 0,0 | eugenia caryophyllata | 0,0 | -0,5 |
| salvia officinalis | 0,7 | 0,0 | fumaria officinalis | 0,0 | -0,5 |
| nasturtium officinalis | 0,6 | -0,1 | gentiana lutea | 0,0 | -0,5 |
| arctium lappa | 0,5 | -0,2 | harpagophytum proc | 0,0 | -0,5 |
| berberis vulgaris | 0,5 | -0,2 | hypericum perforatum | 0,0 | -0,5 |
| lithospermum | 0,5 | -0,2 | lespedeza capitata | 0,0 | -0,5 |
| lotus corniculatus | 0,5 | -0,2 | melissa officinalis | 0,0 | -0,5 |
| spiruline nebl | 0,5 | -0,2 | orthosiphon stamineus | 0,0 | -0,5 |
| vaccinium myrtillus | 0,5 | -0,2 | ginseng | 0,0 | -0,5 |
| allium sativum | 0,4 | -0,2 | plantago lanceolata | 0,0 | -0,5 |
| althea officinalis | 0,4 | -0,2 | raphanus sativus niger | 0,0 | -0,5 |
| chrysanthellum amer | 0,4 | -0,2 | rheum officinale | 0,0 | -0,5 |
| fucus vesiculosus | 0,4 | -0,2 | ruscus aculeatus | 0,0 | -0,5 |
| humulus lupulus | 0,4 | -0,2 | salix alba | 0,0 | -0,5 |
| melilotus officinalis | 0,4 | -0,2 | sarsaparilla | 0,0 | -0,5 |
| passiflora incarnata | 0,4 | -0,2 | tilia cordata | 0,0 | -0,5 |
| phaseolus vulgaris | 0,4 | -0,2 | urtica dioica | 0,0 | -0,5 |
| rosa canina | 0,4 | -0,2 | vinca minor | 0,0 | -0,5 |
| spiruline poud | 0,4 | -0,2 | viola tricolor | 0,0 | -0,5 |
| zingiber officinalis | 0,4 | -0,2 | viscum album | 0,0 | -0,5 |
| carduus marianus | 0,3 | -0,3 | agnus castus | 0,0 | -0,5 |

# Magnesium

| | | | | | |
|---|---|---|---|---|---|
| acorus calamus | 2110,0 | 0,0 | hyoscyamus niger | 3063,0 | 0,5 |
| aesculus hippocastanum | 824,2 | -0,7 | hypericum perforatum | 237,2 | -1,0 |
| agnus castus | 1098,0 | -0,6 | juniperus communis | 464,3 | -0,9 |
| alchemilla vulgaris | 2700,0 | 0,3 | lespedeza capitata | 1347,0 | -0,4 |
| allium sativum | 953,1 | -0,6 | lithospermum | 1826,0 | -0,2 |
| althea officinalis | 2202,0 | 0,0 | lotus corniculatus | 5990,0 | 1,9 |
| ananassa comosus | 601,0 | -0,8 | marrubium vulgare | 2046,0 | -0,1 |
| angelica archangelica | 491,6 | -0,9 | medicago sativa | 1617,0 | -0,3 |
| arctium lappa | 1342,0 | -0,4 | melilotus officinalis | 3829,0 | 0,8 |
| ballota foetida | 3065,0 | 0,5 | melissa officinalis | 305,9 | -1,0 |
| berberis vulgaris | 1602,0 | -0,3 | mentha piperita | 3474,0 | 0,7 |
| beta vulgaris | 1164,0 | -0,5 | millefolium | 1201,0 | -0,5 |
| betula alba | 5195,0 | 1,5 | nasturtium officinalis | 5314,0 | 1,6 |
| boldo fragrans | 456,5 | -0,9 | olea europaea | 1297,0 | -0,5 |
| carduus marianus | 22,5 | -1,1 | orthosiphon stamineus | 447,4 | -0,9 |
| carica papaya | 2320,0 | 0,1 | passiflora incarnata | 2976,0 | 0,4 |
| caroube | 797,0 | -0,7 | phaseolus vulgaris | 8270,0 | 3,1 |
| carragheen | 3327,0 | 0,6 | pinus maritima | 2766,0 | 0,3 |
| carum carvi | 707,1 | -0,8 | piper methysticum | 17,2 | -1,1 |
| chamomilla | 1061,0 | -0,6 | plantago lanceolata | 955,9 | -0,6 |
| china officinalis | 50,6 | -1,1 | raphanus sativus niger | 1308,0 | -0,4 |
| chrysanthellum amer | 2708,0 | 0,3 | rhamnus frangula | 792,2 | -0,7 |
| chrysanthemum part | 480,1 | -0,9 | rhamnus purshiana | 956,4 | -0,6 |
| crataegus oxyacantha | 2081,0 | -0,1 | rheum officinale | 1448,0 | -0,4 |
| cupressus semp | 1987,0 | -0,1 | ribes nigrum folia | 3152,0 | 0,5 |
| curcuma | 1868,0 | -0,2 | ribes nigrum fructus | 344,1 | -0,9 |
| cynara scolymus | 456,6 | -0,9 | rosa canina | 736,5 | -0,7 |
| echinacea angustifolia | 1597,0 | -0,3 | rosmarinus officinalis | 3004,0 | 0,4 |
| echinacea purpurea | 3040,0 | 0,4 | ruscus aculeatus | 228,2 | -1,0 |
| eleutherococcus | 781,2 | -0,7 | salix alba | 2824,0 | 0,3 |
| equisetum arvense | 3539,0 | 0,7 | salvia officinalis | 3153,0 | 0,5 |
| erigeron canadensis | 413,6 | -0,9 | sarsaparilla | 1225,0 | -0,5 |
| eschscholtzia cali | 4591,0 | 1,2 | senna | 1523,0 | -0,3 |
| eucalyptus | 1236,0 | -0,5 | solidago virga aurea | 3354,0 | 0,6 |
| eugenia caryophyllata | 184,4 | -1,0 | spiraea ulmaria | 3209,0 | 0,5 |
| fenugrec | 512,0 | -0,9 | spiruline nebl | 823,7 | -0,7 |
| foeniculum vulgare | 4327,0 | 1,1 | spiruline poud | 2506,0 | 0,2 |
| fraxinus excelior | 6586,0 | 2,2 | taraxacum | 3581,0 | 0,7 |
| fucus vesiculosus | 2657,0 | 0,2 | thymus vulgaris | 3152,0 | 0,5 |
| fumaria officinalis | 1518,0 | -0,3 | tilia alburnum | 850,4 | -0,7 |
| gentiana lutea | 1594,0 | -0,3 | tilia cordata | 1080,0 | -0,6 |
| ginkgo biloba | 7326,0 | 2,6 | urtica dioica | 6500,0 | 2,2 |
| ginseng | 250,9 | -1,0 | uva ursi | 999,6 | -0,6 |
| glycyrrhiza glabra | 4076,0 | 1,0 | vaccinium myrtillus | 230,8 | -1,0 |
| hamamelis virginia | 1059,0 | -0,6 | valeriana officinalis | 246,4 | -1,0 |
| harpagophytum proc | 1057,0 | -0,6 | vinca minor | 1595,0 | -0,3 |
| hibiscus sabdariffa | 4500,0 | 1,2 | viola tricolor | 6278,0 | 2,1 |
| hieracium pilosella | 2060,0 | -0,1 | viscum album | 3721,0 | 0,8 |
| humulus lupulus | 47,7 | -1,1 | vitis vinifera | 10910,0 | 4,4 |
| hydrocotyle asiatica | 2227,0 | 0,0 | zingiber officinalis | 3840,0 | 0,8 |

# Magnesium

| | | | | | |
|---|---|---|---|---|---|
| vitis vinifera | 10910,0 | 4,4 | vinca minor | 1595,0 | -0,3 |
| phaseolus vulgaris | 8270,0 | 3,1 | senna | 1523,0 | -0,3 |
| ginkgo biloba | 7326,0 | 2,6 | fumaria officinalis | 1518,0 | -0,3 |
| fraxinus excelior | 6586,0 | 2,2 | rheum officinale | 1448,0 | -0,4 |
| urtica dioica | 6500,0 | 2,2 | arctium lappa | 1342,0 | -0,4 |
| viola tricolor | 6278,0 | 2,1 | lespedeza capitata | 1347,0 | -0,4 |
| lotus corniculatus | 5990,0 | 1,9 | raphanus sativus niger | 1308,0 | -0,4 |
| nasturtium officinalis | 5314,0 | 1,6 | olea europaea | 1297,0 | -0,5 |
| betula alba | 5195,0 | 1,5 | eucalyptus | 1236,0 | -0,5 |
| eschscholtzia cali | 4591,0 | 1,2 | sarsaparilla | 1225,0 | -0,5 |
| hibiscus sabdariffa | 4500,0 | 1,2 | millefolium | 1201,0 | -0,5 |
| foeniculum vulgare | 4327,0 | 1,1 | beta vulgaris | 1164,0 | -0,5 |
| glycyrrhiza glabra | 4076,0 | 1,0 | agnus castus | 1098,0 | -0,6 |
| melilotus officinalis | 3829,0 | 0,8 | tilia cordata | 1080,0 | -0,6 |
| zingiber officinalis | 3840,0 | 0,8 | hamamelis virginia | 1059,0 | -0,6 |
| viscum album | 3721,0 | 0,8 | harpagophytum proc | 1057,0 | -0,6 |
| taraxacum | 3581,0 | 0,7 | chamomilla | 1061,0 | -0,6 |
| equisetum arvense | 3539,0 | 0,7 | uva ursi | 999,6 | -0,6 |
| mentha piperita | 3474,0 | 0,7 | allium sativum | 953,1 | -0,6 |
| solidago virga aurea | 3354,0 | 0,6 | plantago lanceolata | 955,9 | -0,6 |
| carragheen | 3327,0 | 0,6 | rhamnus purshiana | 956,4 | -0,6 |
| spiraea ulmaria | 3209,0 | 0,5 | tilia alburnum | 850,4 | -0,7 |
| ribes nigrum folia | 3152,0 | 0,5 | aesculus hippocastanum | 824,2 | -0,7 |
| salvia officinalis | 3153,0 | 0,5 | spiruline nebl | 823,7 | -0,7 |
| thymus vulgaris | 3152,0 | 0,5 | caroube | 797,0 | -0,7 |
| ballota foetida | 3065,0 | 0,5 | rhamnus frangula | 792,2 | -0,7 |
| hyoscyamus niger | 3063,0 | 0,5 | eleutherococcus | 781,2 | -0,7 |
| echinacea purpurea | 3040,0 | 0,4 | rosa canina | 736,5 | -0,7 |
| rosmarinus officinalis | 3004,0 | 0,4 | carum carvi | 707,1 | -0,8 |
| passiflora incarnata | 2976,0 | 0,4 | ananassa comosus | 601,0 | -0,8 |
| salix alba | 2824,0 | 0,3 | fenugrec | 512,0 | -0,9 |
| pinus maritima | 2766,0 | 0,3 | angelica archangelica | 491,6 | -0,9 |
| chrysanthellum amer | 2708,0 | 0,3 | chrysanthemum part | 480,1 | -0,9 |
| alchemilla vulgaris | 2700,0 | 0,3 | cynara scolymus | 456,6 | -0,9 |
| fucus vesiculosus | 2657,0 | 0,2 | boldo fragrans | 456,5 | -0,9 |
| spiruline poud | 2506,0 | 0,2 | juniperus communis | 464,3 | -0,9 |
| carica papaya | 2320,0 | 0,1 | orthosiphon stamineus | 447,4 | -0,9 |
| hydrocotyle asiatica | 2227,0 | 0,0 | erigeron canadensis | 413,6 | -0,9 |
| althea officinalis | 2202,0 | 0,0 | ribes nigrum fructus | 344,1 | -0,9 |
| acorus calamus | 2110,0 | 0,0 | melissa officinalis | 305,9 | -1,0 |
| crataegus oxyacantha | 2081,0 | -0,1 | ginseng | 250,9 | -1,0 |
| hieracium pilosella | 2060,0 | -0,1 | valeriana officinalis | 246,4 | -1,0 |
| marrubium vulgare | 2046,0 | -0,1 | hypericum perforatum | 237,2 | -1,0 |
| cupressus semp | 1987,0 | -0,1 | ruscus aculeatus | 228,2 | -1,0 |
| curcuma | 1868,0 | -0,2 | vaccinium myrtillus | 230,8 | -1,0 |
| lithospermum | 1826,0 | -0,2 | eugenia caryophyllata | 184,4 | -1,0 |
| berberis vulgaris | 1602,0 | -0,3 | humulus lupulus | 47,7 | -1,1 |
| medicago sativa | 1617,0 | -0,3 | china officinalis | 50,6 | -1,1 |
| echinacea angustifolia | 1597,0 | -0,3 | carduus marianus | 22,5 | -1,1 |
| gentiana lutea | 1594,0 | -0,3 | piper methysticum | 17,2 | -1,1 |

# Manganese

| | | | | | |
|---|---|---|---|---|---|
| acorus calamus | 85,4 | 0,1 | hyoscyamus niger | 15,6 | -0,3 |
| aesculus hippocastanum | 1,6 | -0,4 | hypericum perforatum | 6,1 | -0,4 |
| agnus castus | 5,5 | -0,4 | juniperus communis | 32,3 | -0,2 |
| alchemilla vulgaris | 58,1 | 0,0 | lespedeza capitata | 23,9 | -0,3 |
| allium sativum | 12,2 | -0,4 | lithospermum | 29,0 | -0,2 |
| althea officinalis | 7,4 | -0,4 | lotus corniculatus | 60,5 | 0,0 |
| ananassa comosus | 81,5 | 0,1 | marrubium vulgare | 15,1 | -0,3 |
| angelica archangelica | 4,7 | -0,4 | medicago sativa | 11,9 | -0,4 |
| arctium lappa | 20,5 | -0,3 | melilotus officinalis | 17,0 | -0,3 |
| ballota foetida | 51,1 | -0,1 | melissa officinalis | 2,5 | -0,4 |
| berberis vulgaris | 28,2 | -0,2 | mentha piperita | 61,1 | 0,0 |
| beta vulgaris | 28,6 | -0,2 | millefolium | 16,9 | -0,3 |
| betula alba | 712,6 | 4,4 | nasturtium officinalis | 70,9 | 0,0 |
| boldo fragrans | 6,7 | -0,4 | olea europaea | 13,3 | -0,3 |
| carduus marianus | 0,2 | -0,4 | orthosiphon stamineus | 4,7 | -0,4 |
| carica papaya | 9,4 | -0,4 | passiflora incarnata | 35,5 | -0,2 |
| caroube | 27,2 | -0,3 | phaseolus vulgaris | 48,2 | -0,1 |
| carragheen | 11,1 | -0,4 | pinus maritima | 291,2 | 1,5 |
| carum carvi | 5,7 | -0,4 | piper methysticum | 0,0 | -0,4 |
| chamomilla | 9,9 | -0,4 | plantago lanceolata | 16,7 | -0,3 |
| china officinalis | 1,3 | -0,4 | raphanus sativus niger | 5,5 | -0,4 |
| chrysanthellum amer | 41,7 | -0,2 | rhamnus frangula | 67,0 | 0,0 |
| chrysanthemum part | 3,6 | -0,4 | rhamnus purshiana | 50,0 | -0,1 |
| crataegus oxyacantha | 6,8 | -0,4 | rheum officinale | 3,5 | -0,4 |
| cupressus semp | 30,5 | -0,2 | ribes nigrum folia | 19,0 | -0,3 |
| curcuma | 60,3 | 0,0 | ribes nigrum fructus | 6,6 | -0,4 |
| cynara scolymus | 4,4 | -0,4 | rosa canina | 14,7 | -0,3 |
| echinacea angustifolia | 3,3 | -0,4 | rosmarinus officinalis | 21,9 | -0,3 |
| echinacea purpurea | 6,6 | -0,4 | ruscus aculeatus | 10,5 | -0,4 |
| eleutherococcus | 7,0 | -0,4 | salix alba | 170,0 | 0,7 |
| equisetum arvense | 13,6 | -0,3 | salvia officinalis | 28,6 | -0,2 |
| erigeron canadensis | 5,9 | -0,4 | sarsaparilla | 18,7 | -0,3 |
| eschscholtzia cali | 120,7 | 0,4 | senna | 3,9 | -0,4 |
| eucalyptus | 909,6 | 5,7 | solidago virga aurea | 37,3 | -0,2 |
| eugenia caryophyllata | 5,8 | -0,4 | spiraea ulmaria | 20,4 | -0,3 |
| fenugrec | 1,5 | -0,4 | spiruline nebl | 5,4 | -0,4 |
| foeniculum vulgare | 7,3 | -0,4 | spiruline poud | 35,3 | -0,2 |
| fraxinus excelior | 60,3 | 0,0 | taraxacum | 19,8 | -0,3 |
| fucus vesiculosus | 8,3 | -0,4 | thymus vulgaris | 59,1 | 0,0 |
| fumaria officinalis | 57,8 | 0,0 | tilia alburnum | 22,5 | -0,3 |
| gentiana lutea | 49,0 | -0,1 | tilia cordata | 5,6 | -0,4 |
| ginkgo biloba | 39,2 | -0,2 | urtica dioica | 27,0 | -0,3 |
| ginseng | 3,2 | -0,4 | uva ursi | 12,8 | -0,4 |
| glycyrrhiza glabra | 14,4 | -0,3 | vaccinium myrtillus | 36,8 | -0,2 |
| hamamelis virginia | 141,0 | 0,5 | valeriana officinalis | 8,3 | -0,4 |
| harpagophytum proc | 8,7 | -0,4 | vinca minor | 183,7 | 0,8 |
| hibiscus sabdariffa | 864,4 | 5,4 | viola tricolor | 190,7 | 0,9 |
| hieracium pilosella | 81,3 | 0,1 | viscum album | 273,0 | 1,4 |
| humulus lupulus | 0,2 | -0,4 | vitis vinifera | 268,3 | 1,4 |
| hydrocotyle asiatica | 93,4 | 0,2 | zingiber officinalis | 217,0 | 1,0 |

# Manganese

| | | | | | |
|---|---|---|---|---|---|
| eucalyptus | 909,6 | 5,7 | sarsaparilla | 18,7 | -0,3 |
| hibiscus sabdariffa | 864,4 | 5,4 | millefolium | 16,9 | -0,3 |
| betula alba | 712,6 | 4,4 | melilotus officinalis | 17,0 | -0,3 |
| pinus maritima | 291,2 | 1,5 | plantago lanceolata | 16,7 | -0,3 |
| viscum album | 273,0 | 1,4 | hyoscyamus niger | 15,6 | -0,3 |
| vitis vinifera | 268,3 | 1,4 | marrubium vulgare | 15,1 | -0,3 |
| zingiber officinalis | 217,0 | 1,0 | rosa canina | 14,7 | -0,3 |
| viola tricolor | 190,7 | 0,9 | olea europaea | 13,3 | -0,3 |
| vinca minor | 183,7 | 0,8 | equisetum arvense | 13,6 | -0,3 |
| salix alba | 170,0 | 0,7 | glycyrrhiza glabra | 14,4 | -0,3 |
| hamamelis virginia | 141,0 | 0,5 | allium sativum | 12,2 | -0,4 |
| eschscholtzia cali | 120,7 | 0,4 | uva ursi | 12,8 | -0,4 |
| hydrocotyle asiatica | 93,4 | 0,2 | medicago sativa | 11,9 | -0,4 |
| acorus calamus | 85,4 | 0,1 | carragheen | 11,1 | -0,4 |
| ananassa comosus | 81,5 | 0,1 | ruscus aculeatus | 10,5 | -0,4 |
| hieracium pilosella | 81,3 | 0,1 | chamomilla | 9,9 | -0,4 |
| nasturtium officinalis | 70,9 | 0,0 | carica papaya | 9,4 | -0,4 |
| rhamnus frangula | 67,0 | 0,0 | althea officinalis | 7,4 | -0,4 |
| mentha piperita | 61,1 | 0,0 | foeniculum vulgare | 7,3 | -0,4 |
| curcuma | 60,3 | 0,0 | fucus vesiculosus | 8,3 | -0,4 |
| fraxinus excelior | 60,3 | 0,0 | harpagophytum proc | 8,7 | -0,4 |
| lotus corniculatus | 60,5 | 0,0 | valeriana officinalis | 8,3 | -0,4 |
| alchemilla vulgaris | 58,1 | 0,0 | crataegus oxyacantha | 6,8 | -0,4 |
| fumaria officinalis | 57,8 | 0,0 | echinacea purpurea | 6,6 | -0,4 |
| thymus vulgaris | 59,1 | 0,0 | eleutherococcus | 7,0 | -0,4 |
| ballota foetida | 51,1 | -0,1 | erigeron canadensis | 5,9 | -0,4 |
| gentiana lutea | 49,0 | -0,1 | eugenia caryophyllata | 5,8 | -0,4 |
| rhamnus purshiana | 50,0 | -0,1 | boldo fragrans | 6,7 | -0,4 |
| phaseolus vulgaris | 48,2 | -0,1 | hypericum perforatum | 6,1 | -0,4 |
| chrysanthellum amer | 41,7 | -0,2 | ribes nigrum fructus | 6,6 | -0,4 |
| ginkgo biloba | 39,2 | -0,2 | angelica archangelica | 4,7 | -0,4 |
| solidago virga aurea | 37,3 | -0,2 | carum carvi | 5,7 | -0,4 |
| passiflora incarnata | 35,5 | -0,2 | cynara scolymus | 4,4 | -0,4 |
| vaccinium myrtillus | 36,8 | -0,2 | orthosiphon stamineus | 4,7 | -0,4 |
| spiruline poud | 35,3 | -0,2 | raphanus sativus niger | 5,5 | -0,4 |
| juniperus communis | 32,3 | -0,2 | spiruline nebl | 5,4 | -0,4 |
| cupressus semp | 30,5 | -0,2 | tilia cordata | 5,6 | -0,4 |
| berberis vulgaris | 28,2 | -0,2 | agnus castus | 5,5 | -0,4 |
| beta vulgaris | 28,6 | -0,2 | senna | 3,9 | -0,4 |
| lithospermum | 29,0 | -0,2 | chrysanthemum part | 3,6 | -0,4 |
| salvia officinalis | 28,6 | -0,2 | echinacea angustifolia | 3,3 | -0,4 |
| caroube | 27,2 | -0,3 | ginseng | 3,2 | -0,4 |
| urtica dioica | 27,0 | -0,3 | rheum officinale | 3,5 | -0,4 |
| lespedeza capitata | 23,9 | -0,3 | aesculus hippocastanum | 1,6 | -0,4 |
| tilia alburnum | 22,5 | -0,3 | fenugrec | 1,5 | -0,4 |
| rosmarinus officinalis | 21,9 | -0,3 | melissa officinalis | 2,5 | -0,4 |
| arctium lappa | 20,5 | -0,3 | carduus marianus | 0,2 | -0,4 |
| spiraea ulmaria | 20,4 | -0,3 | humulus lupulus | 0,2 | -0,4 |
| taraxacum | 19,8 | -0,3 | piper methysticum | 0,0 | -0,4 |
| ribes nigrum folia | 19,0 | -0,3 | china officinalis | 1,3 | -0,4 |

# Mercury

| | | | | | |
|---|---|---|---|---|---|
| acorus calamus | 0,0 | -0,4 | hyoscyamus niger | 0,0 | -0,4 |
| aesculus hippocastanum | 0,0 | -0,4 | hypericum perforatum | 0,7 | 1,0 |
| agnus castus | 0,0 | -0,4 | juniperus communis | 0,0 | -0,4 |
| alchemilla vulgaris | 0,0 | -0,4 | lespedeza capitata | 0,0 | -0,4 |
| allium sativum | 0,0 | -0,4 | lithospermum | 0,0 | -0,4 |
| althea officinalis | 0,0 | -0,4 | lotus corniculatus | 0,0 | -0,4 |
| ananassa comosus | 0,0 | -0,4 | marrubium vulgare | 0,8 | 1,2 |
| angelica archangelica | 2,0 | 3,6 | medicago sativa | 0,4 | 0,4 |
| arctium lappa | 0,0 | -0,4 | melilotus officinalis | 0,0 | -0,4 |
| ballota foetida | 0,0 | -0,4 | melissa officinalis | 0,0 | -0,4 |
| berberis vulgaris | 0,0 | -0,4 | mentha piperita | 0,0 | -0,4 |
| beta vulgaris | 0,5 | 0,6 | millefolium | 0,0 | -0,4 |
| betula alba | 0,0 | -0,4 | nasturtium officinalis | 1,2 | 2,0 |
| boldo fragrans | 0,1 | -0,3 | olea europaea | 0,1 | -0,2 |
| carduus marianus | 0,0 | -0,4 | orthosiphon stamineus | 0,0 | -0,4 |
| carica papaya | 1,6 | 2,8 | passiflora incarnata | 0,0 | -0,4 |
| caroube | 0,7 | 1,0 | phaseolus vulgaris | 0,0 | -0,4 |
| carragheen | 0,0 | -0,4 | pinus maritima | 0,0 | -0,4 |
| carum carvi | 0,0 | -0,4 | piper methysticum | 0,0 | -0,4 |
| chamomilla | 0,2 | 0,0 | plantago lanceolata | 0,0 | -0,4 |
| china officinalis | 0,8 | 1,2 | raphanus sativus niger | 0,0 | -0,4 |
| chrysanthellum amer | 0,0 | -0,4 | rhamnus frangula | 1,6 | 2,8 |
| chrysanthemum part | 0,1 | -0,3 | rhamnus purshiana | 0,0 | -0,4 |
| crataegus oxyacantha | 0,3 | 0,2 | rheum officinale | 0,0 | -0,4 |
| cupressus semp | 0,0 | -0,4 | ribes nigrum folia | 0,6 | 0,8 |
| curcuma | 0,0 | -0,4 | ribes nigrum fructus | 0,0 | -0,4 |
| cynara scolymus | 0,0 | -0,4 | rosa canina | 0,0 | -0,4 |
| echinacea angustifolia | 0,0 | -0,4 | rosmarinus officinalis | 0,6 | 0,8 |
| echinacea purpurea | 0,8 | 1,2 | ruscus aculeatus | 0,0 | -0,4 |
| eleutherococcus | 0,3 | 0,2 | salix alba | 0,0 | -0,4 |
| equisetum arvense | 0,0 | -0,4 | salvia officinalis | 0,0 | -0,4 |
| erigeron canadensis | 0,2 | 0,0 | sarsaparilla | 0,0 | -0,4 |
| eschscholtzia cali | 0,4 | 0,4 | senna | 0,0 | -0,4 |
| eucalyptus | 0,0 | -0,4 | solidago virga aurea | 0,6 | 0,8 |
| eugenia caryophyllata | 0,0 | -0,4 | spiraea ulmaria | 0,1 | -0,2 |
| fenugrec | 0,0 | -0,4 | spiruline nebl | 0,0 | -0,4 |
| foeniculum vulgare | 0,0 | -0,4 | spiruline poud | 0,0 | -0,4 |
| fraxinus excelior | 2,5 | 4,6 | taraxacum | 0,6 | 0,8 |
| fucus vesiculosus | 0,0 | -0,4 | thymus vulgaris | 0,0 | -0,4 |
| fumaria officinalis | 0,0 | -0,4 | tilia alburnum | 0,0 | -0,4 |
| gentiana lutea | 0,0 | -0,4 | tilia cordata | 0,0 | -0,4 |
| ginkgo biloba | 0,0 | -0,4 | urtica dioica | 0,1 | -0,2 |
| ginseng | 0,0 | -0,4 | uva ursi | 0,0 | -0,4 |
| glycyrrhiza glabra | 0,0 | -0,4 | vaccinium myrtillus | 0,8 | 1,2 |
| hamamelis virginia | 0,0 | -0,4 | valeriana officinalis | 0,0 | -0,4 |
| harpagophytum proc | 0,0 | -0,4 | vinca minor | 0,0 | -0,4 |
| hibiscus sabdariffa | 0,0 | -0,4 | viola tricolor | 0,0 | -0,4 |
| hieracium pilosella | 2,4 | 4,4 | viscum album | 0,0 | -0,4 |
| humulus lupulus | 0,0 | -0,4 | vitis vinifera | 0,0 | -0,4 |
| hydrocotyle asiatica | 1,0 | 1,6 | zingiber officinalis | 0,0 | -0,4 |

# Mercury

| | | | | | |
|---|---|---|---|---|---|
| fraxinus excelior | 2,5 | 4,6 | equisetum arvense | 0,0 | -0,4 |
| hieracium pilosella | 2,4 | 4,4 | eucalyptus | 0,0 | -0,4 |
| angelica archangelica | 2,0 | 3,6 | eugenia caryophyllata | 0,0 | -0,4 |
| carica papaya | 1,6 | 2,8 | fenugrec | 0,0 | -0,4 |
| rhamnus frangula | 1,6 | 2,8 | foeniculum vulgare | 0,0 | -0,4 |
| nasturtium officinalis | 1,2 | 2,0 | fucus vesiculosus | 0,0 | -0,4 |
| hydrocotyle asiatica | 1,0 | 1,6 | fumaria officinalis | 0,0 | -0,4 |
| echinacea purpurea | 0,8 | 1,2 | gentiana lutea | 0,0 | -0,4 |
| marrubium vulgare | 0,8 | 1,2 | ginkgo biloba | 0,0 | -0,4 |
| china officinalis | 0,8 | 1,2 | glycyrrhiza glabra | 0,0 | -0,4 |
| vaccinium myrtillus | 0,8 | 1,2 | hibiscus sabdariffa | 0,0 | -0,4 |
| caroube | 0,7 | 1,0 | hamamelis virginia | 0,0 | -0,4 |
| hypericum perforatum | 0,7 | 1,0 | harpagophytum proc | 0,0 | -0,4 |
| ribes nigrum folia | 0,6 | 0,8 | humulus lupulus | 0,0 | -0,4 |
| rosmarinus officinalis | 0,6 | 0,8 | hyoscyamus niger | 0,0 | -0,4 |
| solidago virga aurea | 0,6 | 0,8 | juniperus communis | 0,0 | -0,4 |
| taraxacum | 0,6 | 0,8 | lespedeza capitata | 0,0 | -0,4 |
| beta vulgaris | 0,5 | 0,6 | lithospermum | 0,0 | -0,4 |
| eschscholtzia cali | 0,4 | 0,4 | lotus corniculatus | 0,0 | -0,4 |
| medicago sativa | 0,4 | 0,4 | melilotus officinalis | 0,0 | -0,4 |
| crataegus oxyacantha | 0,3 | 0,2 | melissa officinalis | 0,0 | -0,4 |
| eleutherococcus | 0,3 | 0,2 | mentha piperita | 0,0 | -0,4 |
| erigeron canadensis | 0,2 | 0,0 | orthosiphon stamineus | 0,0 | -0,4 |
| chamomilla | 0,2 | 0,0 | ginseng | 0,0 | -0,4 |
| olea europaea | 0,1 | -0,2 | passiflora incarnata | 0,0 | -0,4 |
| spiraea ulmaria | 0,1 | -0,2 | phaseolus vulgaris | 0,0 | -0,4 |
| urtica dioica | 0,1 | -0,2 | pinus maritima | 0,0 | -0,4 |
| chrysanthemum part | 0,1 | -0,3 | piper methysticum | 0,0 | -0,4 |
| boldo fragrans | 0,1 | -0,3 | plantago lanceolata | 0,0 | -0,4 |
| millefolium | 0,0 | -0,4 | raphanus sativus niger | 0,0 | -0,4 |
| acorus calamus | 0,0 | -0,4 | rhamnus purshiana | 0,0 | -0,4 |
| aesculus hippocastanum | 0,0 | -0,4 | rheum officinale | 0,0 | -0,4 |
| alchemilla vulgaris | 0,0 | -0,4 | ribes nigrum fructus | 0,0 | -0,4 |
| allium sativum | 0,0 | -0,4 | rosa canina | 0,0 | -0,4 |
| althea officinalis | 0,0 | -0,4 | ruscus aculeatus | 0,0 | -0,4 |
| ananassa comosus | 0,0 | -0,4 | salix alba | 0,0 | -0,4 |
| arctium lappa | 0,0 | -0,4 | salvia officinalis | 0,0 | -0,4 |
| uva ursi | 0,0 | -0,4 | sarsaparilla | 0,0 | -0,4 |
| ballota foetida | 0,0 | -0,4 | spiruline nebl | 0,0 | -0,4 |
| berberis vulgaris | 0,0 | -0,4 | spiruline poud | 0,0 | -0,4 |
| betula alba | 0,0 | -0,4 | thymus vulgaris | 0,0 | -0,4 |
| carduus marianus | 0,0 | -0,4 | tilia alburnum | 0,0 | -0,4 |
| carragheen | 0,0 | -0,4 | tilia cordata | 0,0 | -0,4 |
| carum carvi | 0,0 | -0,4 | valeriana officinalis | 0,0 | -0,4 |
| senna | 0,0 | -0,4 | vinca minor | 0,0 | -0,4 |
| chrysanthellum amer | 0,0 | -0,4 | viola tricolor | 0,0 | -0,4 |
| cupressus semp | 0,0 | -0,4 | viscum album | 0,0 | -0,4 |
| curcuma | 0,0 | -0,4 | agnus castus | 0,0 | -0,4 |
| cynara scolymus | 0,0 | -0,4 | vitis vinifera | 0,0 | -0,4 |
| echinacea angustifolia | 0,0 | -0,4 | zingiber officinalis | 0,0 | -0,4 |

# Molybdenum

| | | | | | |
|---|---|---|---|---|---|
| acorus calamus | 0,0 | -0,4 | hyoscyamus niger | 1,4 | 1,7 |
| aesculus hippocastanum | 0,0 | -0,4 | hypericum perforatum | 0,0 | -0,4 |
| agnus castus | 0,0 | -0,4 | juniperus communis | 0,0 | -0,4 |
| alchemilla vulgaris | 0,0 | -0,4 | lespedeza capitata | 0,5 | 0,4 |
| allium sativum | 0,0 | -0,4 | lithospermum | 0,0 | -0,4 |
| althea officinalis | 2,2 | 2,8 | lotus corniculatus | 0,0 | -0,4 |
| ananassa comosus | 0,0 | -0,4 | marrubium vulgare | 0,5 | 0,4 |
| angelica archangelica | 0,0 | -0,4 | medicago sativa | 0,0 | -0,4 |
| arctium lappa | 0,0 | -0,4 | melilotus officinalis | 0,0 | -0,4 |
| ballota foetida | 1,3 | 1,5 | melissa officinalis | 0,0 | -0,4 |
| berberis vulgaris | 0,8 | 0,8 | mentha piperita | 0,0 | -0,4 |
| beta vulgaris | 0,0 | -0,4 | millefolium | 0,0 | -0,4 |
| betula alba | 0,0 | -0,4 | nasturtium officinalis | 0,0 | -0,4 |
| boldo fragrans | 1,0 | 1,1 | olea europaea | 0,0 | -0,4 |
| carduus marianus | 0,0 | -0,4 | orthosiphon stamineus | 0,0 | -0,4 |
| carica papaya | 0,0 | -0,4 | passiflora incarnata | 0,0 | -0,4 |
| caroube | 0,0 | -0,4 | phaseolus vulgaris | 1,2 | 1,4 |
| carragheen | 0,0 | -0,4 | pinus maritima | 0,0 | -0,4 |
| carum carvi | 0,0 | -0,4 | piper methysticum | 0,0 | -0,4 |
| chamomilla | 0,0 | -0,4 | plantago lanceolata | 0,0 | -0,4 |
| china officinalis | 0,0 | -0,4 | raphanus sativus niger | 0,0 | -0,4 |
| chrysanthellum amer | 0,0 | -0,4 | rhamnus frangula | 0,0 | -0,4 |
| chrysanthemum part | 0,0 | -0,4 | rhamnus purshiana | 0,0 | -0,4 |
| crataegus oxyacantha | 0,0 | -0,4 | rheum officinale | 0,0 | -0,4 |
| cupressus semp | 0,0 | -0,4 | ribes nigrum folia | 0,0 | -0,4 |
| curcuma | 0,0 | -0,4 | ribes nigrum fructus | 0,0 | -0,4 |
| cynara scolymus | 0,0 | -0,4 | rosa canina | 0,2 | -0,1 |
| echinacea angustifolia | 0,0 | -0,4 | rosmarinus officinalis | 2,8 | 3,7 |
| echinacea purpurea | 0,0 | -0,4 | ruscus aculeatus | 0,0 | -0,4 |
| eleutherococcus | 0,0 | -0,4 | salix alba | 0,0 | -0,4 |
| equisetum arvense | 0,0 | -0,4 | salvia officinalis | 0,1 | -0,2 |
| erigeron canadensis | 0,0 | -0,4 | sarsaparilla | 3,3 | 4,4 |
| eschscholtzia cali | 0,0 | -0,4 | senna | 0,0 | -0,4 |
| eucalyptus | 0,0 | -0,4 | solidago virga aurea | 0,0 | -0,4 |
| eugenia caryophyllata | 0,0 | -0,4 | spiraea ulmaria | 0,6 | 0,5 |
| fenugrec | 0,0 | -0,4 | spiruline nebl | 0,7 | 0,7 |
| foeniculum vulgare | 1,4 | 1,7 | spiruline poud | 0,0 | -0,4 |
| fraxinus excelior | 0,0 | -0,4 | taraxacum | 0,0 | -0,4 |
| fucus vesiculosus | 0,0 | -0,4 | thymus vulgaris | 0,0 | -0,4 |
| fumaria officinalis | 0,0 | -0,4 | tilia alburnum | 0,0 | -0,4 |
| gentiana lutea | 0,0 | -0,4 | tilia cordata | 0,0 | -0,4 |
| ginkgo biloba | 0,0 | -0,4 | urtica dioica | 0,0 | -0,4 |
| ginseng | 0,0 | -0,4 | uva ursi | 0,0 | -0,4 |
| glycyrrhiza glabra | 1,8 | 2,3 | vaccinium myrtillus | 0,0 | -0,4 |
| hamamelis virginia | 0,0 | -0,4 | valeriana officinalis | 1,0 | 1,1 |
| harpagophytum proc | 0,0 | -0,4 | vinca minor | 0,0 | -0,4 |
| hibiscus sabdariffa | 0,0 | -0,4 | viola tricolor | 0,0 | -0,4 |
| hieracium pilosella | 0,0 | -0,4 | viscum album | 0,0 | -0,4 |
| humulus lupulus | 0,0 | -0,3 | vitis vinifera | 4,0 | 5,4 |
| hydrocotyle asiatica | 0,5 | 0,4 | zingiber officinalis | 0,0 | -0,4 |

# Molybdenum

| | | | | | |
|---|---|---|---|---|---|
| vitis vinifera | 4,0 | 5,4 | eugenia caryophyllata | 0,0 | -0,4 |
| sarsaparilla | 3,3 | 4,4 | fenugrec | 0,0 | -0,4 |
| rosmarinus officinalis | 2,8 | 3,7 | fraxinus excelior | 0,0 | -0,4 |
| althea officinalis | 2,2 | 2,8 | fucus vesiculosus | 0,0 | -0,4 |
| glycyrrhiza glabra | 1,8 | 2,3 | fumaria officinalis | 0,0 | -0,4 |
| foeniculum vulgare | 1,4 | 1,7 | gentiana lutea | 0,0 | -0,4 |
| hyoscyamus niger | 1,4 | 1,7 | ginkgo biloba | 0,0 | -0,4 |
| ballota foetida | 1,3 | 1,5 | hibiscus sabdariffa | 0,0 | -0,4 |
| phaseolus vulgaris | 1,2 | 1,4 | hamamelis virginia | 0,0 | -0,4 |
| boldo fragrans | 1,0 | 1,1 | harpagophytum proc | 0,0 | -0,4 |
| valeriana officinalis | 1,0 | 1,1 | hieracium pilosella | 0,0 | -0,4 |
| berberis vulgaris | 0,8 | 0,8 | hypericum perforatum | 0,0 | -0,4 |
| spiruline nebl | 0,7 | 0,7 | juniperus communis | 0,0 | -0,4 |
| spiraea ulmaria | 0,6 | 0,5 | lithospermum | 0,0 | -0,4 |
| hydrocotyle asiatica | 0,5 | 0,4 | lotus corniculatus | 0,0 | -0,4 |
| lespedeza capitata | 0,5 | 0,4 | chamomilla | 0,0 | -0,4 |
| marrubium vulgare | 0,5 | 0,4 | medicago sativa | 0,0 | -0,4 |
| rosa canina | 0,2 | -0,1 | melilotus officinalis | 0,0 | -0,4 |
| salvia officinalis | 0,1 | -0,2 | carica papaya | 0,0 | -0,4 |
| humulus lupulus | 0,0 | -0,3 | melissa officinalis | 0,0 | -0,4 |
| millefolium | 0,0 | -0,4 | mentha piperita | 0,0 | -0,4 |
| acorus calamus | 0,0 | -0,4 | nasturtium officinalis | 0,0 | -0,4 |
| aesculus hippocastanum | 0,0 | -0,4 | orthosiphon stamineus | 0,0 | -0,4 |
| alchemilla vulgaris | 0,0 | -0,4 | ginseng | 0,0 | -0,4 |
| allium sativum | 0,0 | -0,4 | passiflora incarnata | 0,0 | -0,4 |
| ananassa comosus | 0,0 | -0,4 | pinus maritima | 0,0 | -0,4 |
| angelica archangelica | 0,0 | -0,4 | piper methysticum | 0,0 | -0,4 |
| arctium lappa | 0,0 | -0,4 | plantago lanceolata | 0,0 | -0,4 |
| uva ursi | 0,0 | -0,4 | china officinalis | 0,0 | -0,4 |
| beta vulgaris | 0,0 | -0,4 | raphanus sativus niger | 0,0 | -0,4 |
| betula alba | 0,0 | -0,4 | rhamnus frangula | 0,0 | -0,4 |
| carduus marianus | 0,0 | -0,4 | rhamnus purshiana | 0,0 | -0,4 |
| caroube | 0,0 | -0,4 | rheum officinale | 0,0 | -0,4 |
| carragheen | 0,0 | -0,4 | ribes nigrum folia | 0,0 | -0,4 |
| carum carvi | 0,0 | -0,4 | ribes nigrum fructus | 0,0 | -0,4 |
| senna | 0,0 | -0,4 | ruscus aculeatus | 0,0 | -0,4 |
| chrysanthellum amer | 0,0 | -0,4 | salix alba | 0,0 | -0,4 |
| chrysanthemum part | 0,0 | -0,4 | solidago virga aurea | 0,0 | -0,4 |
| crataegus oxyacantha | 0,0 | -0,4 | spiruline poud | 0,0 | -0,4 |
| cupressus semp | 0,0 | -0,4 | taraxacum | 0,0 | -0,4 |
| curcuma | 0,0 | -0,4 | thymus vulgaris | 0,0 | -0,4 |
| cynara scolymus | 0,0 | -0,4 | tilia alburnum | 0,0 | -0,4 |
| echinacea angustifolia | 0,0 | -0,4 | tilia cordata | 0,0 | -0,4 |
| echinacea purpurea | 0,0 | -0,4 | urtica dioica | 0,0 | -0,4 |
| olea europaea | 0,0 | -0,4 | vaccinium myrtillus | 0,0 | -0,4 |
| eleutherococcus | 0,0 | -0,4 | vinca minor | 0,0 | -0,4 |
| equisetum arvense | 0,0 | -0,4 | viola tricolor | 0,0 | -0,4 |
| erigeron canadensis | 0,0 | -0,4 | viscum album | 0,0 | -0,4 |
| eschscholtzia cali | 0,0 | -0,4 | agnus castus | 0,0 | -0,4 |
| eucalyptus | 0,0 | -0,4 | zingiber officinalis | 0,0 | -0,4 |

# Nickel

| | | | | | |
|---|---|---|---|---|---|
| acorus calamus | 5,8 | 0,8 | hyoscyamus niger | 0,3 | -0,6 |
| aesculus hippocastanum | 0,8 | -0,4 | hypericum perforatum | 0,6 | -0,5 |
| agnus castus | 0,1 | -0,6 | juniperus communis | 2,9 | 0,1 |
| alchemilla vulgaris | 3,3 | 0,2 | lespedeza capitata | 4,4 | 0,5 |
| allium sativum | 0,0 | -0,6 | lithospermum | 1,6 | -0,2 |
| althea officinalis | 0,9 | -0,4 | lotus corniculatus | 3,1 | 0,2 |
| ananassa comosus | 1,9 | -0,2 | marrubium vulgare | 0,8 | -0,4 |
| angelica archangelica | 0,3 | -0,6 | medicago sativa | 0,2 | -0,6 |
| arctium lappa | 11,5 | 2,3 | melilotus officinalis | 5,0 | 0,6 |
| ballota foetida | 0,0 | -0,6 | melissa officinalis | 1,2 | -0,3 |
| berberis vulgaris | 0,0 | -0,6 | mentha piperita | 1,3 | -0,3 |
| beta vulgaris | 1,5 | -0,3 | millefolium | 1,5 | -0,3 |
| betula alba | 18,0 | 3,9 | nasturtium officinalis | 3,4 | 0,2 |
| boldo fragrans | 0,0 | -0,6 | olea europaea | 0,8 | -0,4 |
| carduus marianus | 0,9 | -0,4 | orthosiphon stamineus | 0,3 | -0,6 |
| carica papaya | 0,2 | -0,6 | passiflora incarnata | 2,2 | -0,1 |
| caroube | 4,2 | 0,4 | phaseolus vulgaris | 5,6 | 0,8 |
| carragheen | 3,2 | 0,2 | pinus maritima | 17,0 | 3,7 |
| carum carvi | 3,0 | 0,1 | piper methysticum | 0,0 | -0,6 |
| chamomilla | 1,3 | -0,3 | plantago lanceolata | 0,8 | -0,4 |
| china officinalis | 0,0 | -0,6 | raphanus sativus niger | 1,1 | -0,4 |
| chrysanthellum amer | 0,0 | -0,6 | rhamnus frangula | 2,2 | -0,1 |
| chrysanthemum part | 0,1 | -0,6 | rhamnus purshiana | 0,0 | -0,6 |
| crataegus oxyacantha | 7,0 | 1,1 | rheum officinale | 6,1 | 0,9 |
| cupressus semp | 0,0 | -0,6 | ribes nigrum folia | 2,4 | 0,0 |
| curcuma | 0,0 | -0,6 | ribes nigrum fructus | 0,0 | -0,6 |
| cynara scolymus | 0,3 | -0,6 | rosa canina | 0,0 | -0,6 |
| echinacea angustifolia | 1,3 | -0,3 | rosmarinus officinalis | 11,0 | 2,1 |
| echinacea purpurea | 0,4 | -0,5 | ruscus aculeatus | 0,9 | -0,4 |
| eleutherococcus | 1,3 | -0,3 | salix alba | 3,0 | 0,1 |
| equisetum arvense | 0,8 | -0,4 | salvia officinalis | 0,0 | -0,6 |
| erigeron canadensis | 0,1 | -0,6 | sarsaparilla | 0,3 | -0,6 |
| eschscholtzia cali | 0,4 | -0,5 | senna | 3,6 | 0,3 |
| eucalyptus | 3,0 | 0,1 | solidago virga aurea | 1,5 | -0,3 |
| eugenia caryophyllata | 0,0 | -0,6 | spiraea ulmaria | 2,5 | 0,0 |
| fenugrec | 1,6 | -0,2 | spiruline nebl | 0,0 | -0,6 |
| foeniculum vulgare | 2,4 | 0,0 | spiruline poud | 11,2 | 2,2 |
| fraxinus excelior | 2,8 | 0,1 | taraxacum | 1,3 | -0,3 |
| fucus vesiculosus | 0,0 | -0,6 | thymus vulgaris | 0,5 | -0,5 |
| fumaria officinalis | 0,4 | -0,5 | tilia alburnum | 1,9 | -0,2 |
| gentiana lutea | 3,3 | 0,2 | tilia cordata | 0,0 | -0,6 |
| ginkgo biloba | 0,9 | -0,4 | urtica dioica | 1,4 | -0,3 |
| ginseng | 0,7 | -0,5 | uva ursi | 2,0 | -0,1 |
| glycyrrhiza glabra | 0,5 | -0,5 | vaccinium myrtillus | 0,0 | -0,6 |
| hamamelis virginia | 0,0 | -0,6 | valeriana officinalis | 0,0 | -0,6 |
| harpagophytum proc | 0,1 | -0,6 | vinca minor | 6,1 | 0,9 |
| hibiscus sabdariffa | 4,8 | 0,6 | viola tricolor | 0,7 | -0,5 |
| hieracium pilosella | 2,1 | -0,1 | viscum album | 20,5 | 4,5 |
| humulus lupulus | 0,3 | -0,6 | vitis vinifera | 15,9 | 3,4 |
| hydrocotyle asiatica | 4,0 | 0,4 | zingiber officinalis | 6,5 | 1,0 |

# Nickel

| | | | | |
|---|---|---|---|---|
| viscum album | 20,5 | 4,5 | raphanus sativus niger | 1,1 | -0,4 |
| betula alba | 18,0 | 3,9 | althea officinalis | 0,9 | -0,4 |
| pinus maritima | 17,0 | 3,7 | carduus marianus | 0,9 | -0,4 |
| vitis vinifera | 15,9 | 3,4 | ginkgo biloba | 0,9 | -0,4 |
| arctium lappa | 11,5 | 2,3 | ruscus aculeatus | 0,9 | -0,4 |
| spiruline poud | 11,2 | 2,2 | aesculus hippocastanum | 0,8 | -0,4 |
| rosmarinus officinalis | 11,0 | 2,1 | olea europaea | 0,8 | -0,4 |
| crataegus oxyacantha | 7,0 | 1,1 | equisetum arvense | 0,8 | -0,4 |
| zingiber officinalis | 6,5 | 1,0 | marrubium vulgare | 0,8 | -0,4 |
| rheum officinale | 6,1 | 0,9 | plantago lanceolata | 0,8 | -0,4 |
| vinca minor | 6,1 | 0,9 | ginseng | 0,7 | -0,5 |
| acorus calamus | 5,8 | 0,8 | viola tricolor | 0,7 | -0,5 |
| phaseolus vulgaris | 5,6 | 0,8 | hypericum perforatum | 0,6 | -0,5 |
| melilotus officinalis | 5,0 | 0,6 | glycyrrhiza glabra | 0,5 | -0,5 |
| hibiscus sabdariffa | 4,8 | 0,6 | thymus vulgaris | 0,5 | -0,5 |
| lespedeza capitata | 4,4 | 0,5 | echinacea purpurea | 0,4 | -0,5 |
| caroube | 4,2 | 0,4 | eschscholtzia cali | 0,4 | -0,5 |
| hydrocotyle asiatica | 4,0 | 0,4 | fumaria officinalis | 0,4 | -0,5 |
| senna | 3,6 | 0,3 | cynara scolymus | 0,3 | -0,6 |
| nasturtium officinalis | 3,4 | 0,2 | angelica archangelica | 0,3 | -0,6 |
| alchemilla vulgaris | 3,3 | 0,2 | humulus lupulus | 0,3 | -0,6 |
| gentiana lutea | 3,3 | 0,2 | hyoscyamus niger | 0,3 | -0,6 |
| carragheen | 3,2 | 0,2 | orthosiphon stamineus | 0,3 | -0,6 |
| lotus corniculatus | 3,1 | 0,2 | sarsaparilla | 0,3 | -0,6 |
| carum carvi | 3,0 | 0,1 | medicago sativa | 0,2 | -0,6 |
| eucalyptus | 3,0 | 0,1 | carica papaya | 0,2 | -0,6 |
| salix alba | 3,0 | 0,1 | erigeron canadensis | 0,1 | -0,6 |
| juniperus communis | 2,9 | 0,1 | harpagophytum proc | 0,1 | -0,6 |
| fraxinus excelior | 2,8 | 0,1 | agnus castus | 0,1 | -0,6 |
| spiraea ulmaria | 2,5 | 0,0 | chrysanthemum part | 0,1 | -0,6 |
| foeniculum vulgare | 2,4 | 0,0 | allium sativum | 0,0 | -0,6 |
| ribes nigrum folia | 2,4 | 0,0 | ballota foetida | 0,0 | -0,6 |
| passiflora incarnata | 2,2 | -0,1 | berberis vulgaris | 0,0 | -0,6 |
| rhamnus frangula | 2,2 | -0,1 | chrysanthellum amer | 0,0 | -0,6 |
| hieracium pilosella | 2,1 | -0,1 | cupressus semp | 0,0 | -0,6 |
| uva ursi | 2,0 | -0,1 | curcuma | 0,0 | -0,6 |
| ananassa comosus | 1,9 | -0,2 | eugenia caryophyllata | 0,0 | -0,6 |
| tilia alburnum | 1,9 | -0,2 | fucus vesiculosus | 0,0 | -0,6 |
| fenugrec | 1,6 | -0,2 | boldo fragrans | 0,0 | -0,6 |
| lithospermum | 1,6 | -0,2 | hamamelis virginia | 0,0 | -0,6 |
| millefolium | 1,5 | -0,3 | piper methysticum | 0,0 | -0,6 |
| beta vulgaris | 1,5 | -0,3 | china officinalis | 0,0 | -0,6 |
| solidago virga aurea | 1,5 | -0,3 | rhamnus purshiana | 0,0 | -0,6 |
| urtica dioica | 1,4 | -0,3 | ribes nigrum fructus | 0,0 | -0,6 |
| echinacea angustifolia | 1,3 | -0,3 | rosa canina | 0,0 | -0,6 |
| eleutherococcus | 1,3 | -0,3 | salvia officinalis | 0,0 | -0,6 |
| chamomilla | 1,3 | -0,3 | spiruline nebl | 0,0 | -0,6 |
| mentha piperita | 1,3 | -0,3 | tilia cordata | 0,0 | -0,6 |
| taraxacum | 1,3 | -0,3 | vaccinium myrtillus | 0,0 | -0,6 |
| melissa officinalis | 1,2 | -0,3 | valeriana officinalis | 0,0 | -0,6 |

# Phosphorus

| | | | | | |
|---|---|---|---|---|---|
| acorus calamus | 2213,0 | 0,5 | hyoscyamus niger | 965,7 | -0,4 |
| aesculus hippocastanum | 1544,0 | 0,0 | hypericum perforatum | 584,1 | -0,7 |
| agnus castus | 1216,0 | -0,2 | juniperus communis | 397,3 | -0,8 |
| alchemilla vulgaris | 1824,0 | 0,2 | lespedeza capitata | 965,5 | -0,4 |
| allium sativum | 4412,0 | 2,0 | lithospermum | 1517,0 | 0,0 |
| althea officinalis | 1110,0 | -0,3 | lotus corniculatus | 5506,0 | 2,7 |
| ananassa comosus | 1250,0 | -0,2 | marrubium vulgare | 2416,0 | 0,6 |
| angelica archangelica | 1231,0 | -0,2 | medicago sativa | 1547,0 | 0,0 |
| arctium lappa | 1477,0 | -0,1 | melilotus officinalis | 635,0 | -0,6 |
| ballota foetida | 1790,0 | 0,2 | melissa officinalis | 215,1 | -0,9 |
| berberis vulgaris | 731,7 | -0,6 | mentha piperita | 1691,0 | 0,1 |
| beta vulgaris | 2124,0 | 0,4 | millefolium | 1331,0 | -0,2 |
| betula alba | 3052,0 | 1,0 | nasturtium officinalis | 3821,0 | 1,6 |
| boldo fragrans | 843,6 | -0,5 | olea europaea | 561,0 | -0,7 |
| carduus marianus | 73,7 | -1,0 | orthosiphon stamineus | 556,9 | -0,7 |
| carica papaya | 1314,0 | -0,2 | passiflora incarnata | 1863,0 | 0,2 |
| caroube | 438,4 | -0,8 | phaseolus vulgaris | 2103,0 | 0,4 |
| carragheen | 784,5 | -0,5 | pinus maritima | 3051,0 | 1,0 |
| carum carvi | 1502,0 | 0,0 | piper methysticum | 53,7 | -1,0 |
| chamomilla | 902,0 | -0,4 | plantago lanceolata | 864,5 | -0,5 |
| china officinalis | 206,1 | -0,9 | raphanus sativus niger | 5319,0 | 2,6 |
| chrysanthellum amer | 2143,0 | 0,4 | rhamnus frangula | 349,5 | -0,8 |
| chrysanthemum part | 894,7 | -0,5 | rhamnus purshiana | 449,6 | -0,8 |
| crataegus oxyacantha | 1242,0 | -0,2 | rheum officinale | 326,3 | -0,8 |
| cupressus semp | 3545,0 | 1,4 | ribes nigrum folia | 998,1 | -0,4 |
| curcuma | 2037,0 | 0,3 | ribes nigrum fructus | 906,9 | -0,4 |
| cynara scolymus | 540,0 | -0,7 | rosa canina | 358,4 | -0,8 |
| echinacea angustifolia | 3763,0 | 1,5 | rosmarinus officinalis | 1072,0 | -0,3 |
| echinacea purpurea | 1291,0 | -0,2 | ruscus aculeatus | 327,2 | -0,8 |
| eleutherococcus | 942,4 | -0,4 | salix alba | 1916,0 | 0,3 |
| equisetum arvense | 841,5 | -0,5 | salvia officinalis | 801,1 | -0,5 |
| erigeron canadensis | 1028,0 | -0,4 | sarsaparilla | 988,2 | -0,4 |
| eschscholtzia cali | 4436,0 | 2,0 | senna | 771,7 | -0,5 |
| eucalyptus | 391,6 | -0,8 | solidago virga aurea | 2519,0 | 0,7 |
| eugenia caryophyllata | 236,6 | -0,9 | spiraea ulmaria | 1531,0 | 0,0 |
| fenugrec | 439,9 | -0,8 | spiruline nebl | 1154,0 | -0,3 |
| foeniculum vulgare | 1681,0 | 0,1 | spiruline poud | 5976,0 | 3,0 |
| fraxinus excelior | 3580,0 | 1,4 | taraxacum | 630,8 | -0,6 |
| fucus vesiculosus | 588,8 | -0,7 | thymus vulgaris | 1358,0 | -0,1 |
| fumaria officinalis | 976,0 | -0,4 | tilia alburnum | 740,9 | -0,6 |
| gentiana lutea | 672,0 | -0,6 | tilia cordata | 803,0 | -0,5 |
| ginkgo biloba | 3241,0 | 1,2 | urtica dioica | 1859,0 | 0,2 |
| ginseng | 366,7 | -0,8 | uva ursi | 877,9 | -0,5 |
| glycyrrhiza glabra | 463,0 | -0,7 | vaccinium myrtillus | 263,6 | -0,9 |
| hamamelis virginia | 593,1 | -0,7 | valeriana officinalis | 406,0 | -0,8 |
| harpagophytum proc | 207,9 | -0,9 | vinca minor | 1522,0 | 0,0 |
| hibiscus sabdariffa | 951,4 | -0,4 | viola tricolor | 8758,0 | 4,9 |
| hieracium pilosella | 2709,0 | 0,8 | viscum album | 4431,0 | 2,0 |
| humulus lupulus | 622,6 | -0,6 | vitis vinifera | 1706,0 | 0,1 |
| hydrocotyle asiatica | 1024,0 | -0,4 | zingiber officinalis | 4130,0 | 1,8 |

# Phosphorus

| | | | | | |
|---|---|---|---|---|---|
| viola tricolor | 8758,0 | 4,9 | erigeron canadensis | 1028,0 | -0,4 |
| spiruline poud | 5976,0 | 3,0 | ribes nigrum folia | 998,1 | -0,4 |
| lotus corniculatus | 5506,0 | 2,7 | fumaria officinalis | 976,0 | -0,4 |
| raphanus sativus niger | 5319,0 | 2,6 | sarsaparilla | 988,2 | -0,4 |
| eschscholtzia cali | 4436,0 | 2,0 | hyoscyamus niger | 965,7 | -0,4 |
| viscum album | 4431,0 | 2,0 | lespedeza capitata | 965,5 | -0,4 |
| allium sativum | 4412,0 | 2,0 | hibiscus sabdariffa | 951,4 | -0,4 |
| zingiber officinalis | 4130,0 | 1,8 | eleutherococcus | 942,4 | -0,4 |
| nasturtium officinalis | 3821,0 | 1,6 | chamomilla | 902,0 | -0,4 |
| echinacea angustifolia | 3763,0 | 1,5 | ribes nigrum fructus | 906,9 | -0,4 |
| fraxinus excelior | 3580,0 | 1,4 | chrysanthemum part | 894,7 | -0,5 |
| cupressus semp | 3545,0 | 1,4 | uva ursi | 877,9 | -0,5 |
| ginkgo biloba | 3241,0 | 1,2 | plantago lanceolata | 864,5 | -0,5 |
| betula alba | 3052,0 | 1,0 | boldo fragrans | 843,6 | -0,5 |
| pinus maritima | 3051,0 | 1,0 | equisetum arvense | 841,5 | -0,5 |
| hieracium pilosella | 2709,0 | 0,8 | salvia officinalis | 801,1 | -0,5 |
| solidago virga aurea | 2519,0 | 0,7 | tilia cordata | 803,0 | -0,5 |
| marrubium vulgare | 2416,0 | 0,6 | carragheen | 784,5 | -0,5 |
| acorus calamus | 2213,0 | 0,5 | senna | 771,7 | -0,5 |
| chrysanthellum amer | 2143,0 | 0,4 | tilia alburnum | 740,9 | -0,6 |
| beta vulgaris | 2124,0 | 0,4 | berberis vulgaris | 731,7 | -0,6 |
| phaseolus vulgaris | 2103,0 | 0,4 | gentiana lutea | 672,0 | -0,6 |
| curcuma | 2037,0 | 0,3 | melilotus officinalis | 635,0 | -0,6 |
| salix alba | 1916,0 | 0,3 | taraxacum | 630,8 | -0,6 |
| passiflora incarnata | 1863,0 | 0,2 | humulus lupulus | 622,6 | -0,6 |
| urtica dioica | 1859,0 | 0,2 | fucus vesiculosus | 588,8 | -0,7 |
| alchemilla vulgaris | 1824,0 | 0,2 | hamamelis virginia | 593,1 | -0,7 |
| ballota foetida | 1790,0 | 0,2 | hypericum perforatum | 584,1 | -0,7 |
| vitis vinifera | 1706,0 | 0,1 | olea europaea | 561,0 | -0,7 |
| foeniculum vulgare | 1681,0 | 0,1 | orthosiphon stamineus | 556,9 | -0,7 |
| mentha piperita | 1691,0 | 0,1 | cynara scolymus | 540,0 | -0,7 |
| medicago sativa | 1547,0 | 0,0 | glycyrrhiza glabra | 463,0 | -0,7 |
| aesculus hippocastanum | 1544,0 | 0,0 | rhamnus purshiana | 449,6 | -0,8 |
| spiraea ulmaria | 1531,0 | 0,0 | caroube | 438,4 | -0,8 |
| vinca minor | 1522,0 | 0,0 | fenugrec | 439,9 | -0,8 |
| lithospermum | 1517,0 | 0,0 | valeriana officinalis | 406,0 | -0,8 |
| carum carvi | 1502,0 | 0,0 | eucalyptus | 391,6 | -0,8 |
| arctium lappa | 1477,0 | -0,1 | juniperus communis | 397,3 | -0,8 |
| thymus vulgaris | 1358,0 | -0,1 | ginseng | 366,7 | -0,8 |
| millefolium | 1331,0 | -0,2 | rhamnus frangula | 349,5 | -0,8 |
| carica papaya | 1314,0 | -0,2 | rosa canina | 358,4 | -0,8 |
| echinacea purpurea | 1291,0 | -0,2 | rheum officinale | 326,3 | -0,8 |
| ananassa comosus | 1250,0 | -0,2 | ruscus aculeatus | 327,2 | -0,8 |
| crataegus oxyacantha | 1242,0 | -0,2 | vaccinium myrtillus | 263,6 | -0,9 |
| angelica archangelica | 1231,0 | -0,2 | eugenia caryophyllata | 236,6 | -0,9 |
| agnus castus | 1216,0 | -0,2 | melissa officinalis | 215,1 | -0,9 |
| spiruline nebl | 1154,0 | -0,3 | harpagophytum proc | 207,9 | -0,9 |
| althea officinalis | 1110,0 | -0,3 | china officinalis | 206,1 | -0,9 |
| rosmarinus officinalis | 1072,0 | -0,3 | carduus marianus | 73,7 | -1,0 |
| hydrocotyle asiatica | 1024,0 | -0,4 | piper methysticum | 53,7 | -1,0 |

# Potassium

| | | | | | |
|---|---|---|---|---|---|
| acorus calamus | 29220,0 | 0,2 | hyoscyamus niger | 53000,0 | 1,2 |
| aesculus hipp | 16470,0 | -0,4 | hypericum perf | 4897,0 | -0,9 |
| agnus castus | 19780,0 | -0,2 | juniperus communis | 9631,0 | -0,7 |
| alchemilla vulgaris | 23200,0 | -0,1 | lespedeza capitata | 14870,0 | -0,4 |
| allium sativum | 17310,0 | -0,3 | lithospermum | 66620,0 | 1,7 |
| althea officinalis | 80,7 | -1,1 | lotus corniculatus | 84090,0 | 2,5 |
| ananassa comosus | 10060,0 | -0,6 | marrubium vulgare | 36400,0 | 0,5 |
| angelica archangelica | 14530,0 | -0,5 | medicago sativa | 42780,0 | 0,7 |
| arctium lappa | 24720,0 | 0,0 | melilotus officinalis | 33630,0 | 0,4 |
| ballota foetida | 42860,0 | 0,7 | melissa officinalis | 3950,0 | -0,9 |
| berberis vulgaris | 12510,0 | -0,5 | mentha piperita | 27120,0 | 0,1 |
| beta vulgaris | 19170,0 | -0,3 | millefolium | 23440,0 | -0,1 |
| betula alba | 24910,0 | 0,0 | nasturtium offi | 158400,0 | 5,6 |
| boldo fragrans | 7821,0 | -0,7 | olea europaea | 10040,0 | -0,6 |
| carduus marianus | 256,7 | -1,0 | orthosiphon stam | 7040,0 | -0,8 |
| carica papaya | 21620,0 | -0,2 | passiflora incarnata | 32200,0 | 0,3 |
| caroube | 7752,0 | -0,7 | phaseolus vulgaris | 75040,0 | 2,1 |
| carragheen | 14000,0 | -0,5 | pinus maritima | 10460,0 | -0,6 |
| carum carvi | 16450,0 | -0,4 | piper methysticum | 704,6 | -1,0 |
| chamomilla | 18500,0 | -0,3 | plantago lanceolata | 13860,0 | -0,5 |
| china officinalis | 1296,0 | -1,0 | raphanus sativus n | 59930,0 | 1,5 |
| chrysanthellum amer | 33060,0 | 0,3 | rhamnus frangula | 3334,0 | -0,9 |
| chrysanthemum part | 2802,0 | -0,9 | rhamnus purshiana | 7237,0 | -0,8 |
| crataegus oxya | 24810,0 | 0,0 | rheum officinale | 10460,0 | -0,6 |
| cupressus semp | 52760,0 | 1,2 | ribes nigrum folia | 18250,0 | -0,3 |
| curcuma | 49690,0 | 1,0 | ribes nigrum fructus | 9168,0 | -0,7 |
| cynara scolymus | 14310,0 | -0,5 | rosa canina | 7330,0 | -0,8 |
| echinacea angust | 14200,0 | -0,5 | rosmarinus offi | 34290,0 | 0,4 |
| echinacea purpurea | 30340,0 | 0,2 | ruscus aculeatus | 8145,0 | -0,7 |
| eleutherococcus | 8564,0 | -0,7 | salix alba | 34050,0 | 0,4 |
| equisetum arvense | 28470,0 | 0,1 | salvia officinalis | 17940,0 | -0,3 |
| erigeron canadensis | 18840,0 | -0,3 | sarsaparilla | 27100,0 | 0,1 |
| eschscholtzia cali | 85830,0 | 2,5 | senna | 21990,0 | -0,1 |
| eucalyptus | 8740,0 | -0,7 | solidago virga aurea | 44890,0 | 0,8 |
| eugenia caryophyllata | 3255,0 | -0,9 | spiraea ulmaria | 21660,0 | -0,2 |
| fenugrec | 8571,0 | -0,7 | spiruline nebl | 27212,0 | 0,1 |
| foeniculum vulgare | 45360,0 | 0,8 | spiruline poud | 12780,0 | -0,5 |
| fraxinus excelior | 36630,0 | 0,5 | taraxacum | 14100,0 | -0,5 |
| fucus vesiculosus | 19870,0 | -0,2 | thymus vulgaris | 24550,0 | 0,0 |
| fumaria officinalis | 50640,0 | 1,1 | tilia alburnum | 10400,0 | -0,6 |
| gentiana lutea | 5587,0 | -0,8 | tilia cordata | 17680,0 | -0,3 |
| ginkgo biloba | 55360,0 | 1,3 | urtica dioica | 76260,0 | 2,1 |
| ginseng | 3062,0 | -0,9 | uva ursi | 10710,0 | -0,6 |
| glycyrrhiza glabra | 22950,0 | -0,1 | vaccinium myrtillus | 2114,0 | -1,0 |
| hamamelis virginia | 8374,0 | -0,7 | valeriana officinalis | 3467,0 | -0,9 |
| harpagophytum proc | 5785,0 | -0,8 | vinca minor | 28120,0 | 0,1 |
| hibiscus sabdariffa | 40110,0 | 0,6 | viola tricolor | 1033,0 | -1,0 |
| hieracium pilosella | 53870,0 | 1,2 | viscum album | 55160,0 | 1,3 |
| humulus lupulus | 26470,0 | 0,1 | vitis vinifera | 10160,0 | -0,6 |
| hydrocotyle asiatica | 29720,0 | 0,2 | zingiber officinalis | 64020,0 | 1,6 |

# Potassium

| | | | | | |
|---|---|---|---|---|---|
| nasturtium offi | 158400,0 | 5,6 | chamomilla | 18500,0 | -0,3 |
| eschscholtzia cali | 85830,0 | 2,5 | ribes nigrum folia | 18250,0 | -0,3 |
| lotus corniculatus | 84090,0 | 2,5 | salvia officinalis | 17940,0 | -0,3 |
| urtica dioica | 76260,0 | 2,1 | tilia cordata | 17680,0 | -0,3 |
| phaseolus vulgaris | 75040,0 | 2,1 | allium sativum | 17310,0 | -0,3 |
| lithospermum | 66620,0 | 1,7 | aesculus hipp | 16470,0 | -0,4 |
| zingiber officinalis | 64020,0 | 1,6 | carum carvi | 16450,0 | -0,4 |
| raphanus sativus n | 59930,0 | 1,5 | lespedeza capitata | 14870,0 | -0,4 |
| ginkgo biloba | 55360,0 | 1,3 | angelica archangelica | 14530,0 | -0,5 |
| viscum album | 55160,0 | 1,3 | cynara scolymus | 14310,0 | -0,5 |
| hieracium pilosella | 53870,0 | 1,2 | echinacea angust | 14200,0 | -0,5 |
| hyoscyamus niger | 53000,0 | 1,2 | carragheen | 14000,0 | -0,5 |
| cupressus semp | 52760,0 | 1,2 | taraxacum | 14100,0 | -0,5 |
| fumaria officinalis | 50640,0 | 1,1 | plantago lanceolata | 13860,0 | -0,5 |
| curcuma | 49690,0 | 1,0 | spiruline poud | 12780,0 | -0,5 |
| foeniculum vulgare | 45360,0 | 0,8 | berberis vulgaris | 12510,0 | -0,5 |
| solidago virga aurea | 44890,0 | 0,8 | uva ursi | 10710,0 | -0,6 |
| ballota foetida | 42860,0 | 0,7 | pinus maritima | 10460,0 | -0,6 |
| medicago sativa | 42780,0 | 0,7 | rheum officinale | 10460,0 | -0,6 |
| hibiscus sabdariffa | 40110,0 | 0,6 | tilia alburnum | 10400,0 | -0,6 |
| fraxinus excelior | 36630,0 | 0,5 | ananassa comosus | 10060,0 | -0,6 |
| marrubium vulgare | 36400,0 | 0,5 | vitis vinifera | 10160,0 | -0,6 |
| rosmarinus offi | 34290,0 | 0,4 | olea europaea | 10040,0 | -0,6 |
| salix alba | 34050,0 | 0,4 | juniperus communis | 9631,0 | -0,7 |
| melilotus officinalis | 33630,0 | 0,4 | ribes nigrum fructus | 9168,0 | -0,7 |
| chrysanthellum amer | 33060,0 | 0,3 | eucalyptus | 8740,0 | -0,7 |
| passiflora incarnata | 32200,0 | 0,3 | eleutherococcus | 8564,0 | -0,7 |
| echinacea purpurea | 30340,0 | 0,2 | fenugrec | 8571,0 | -0,7 |
| hydrocotyle asiatica | 29720,0 | 0,2 | hamamelis virginia | 8374,0 | -0,7 |
| acorus calamus | 29220,0 | 0,2 | ruscus aculeatus | 8145,0 | -0,7 |
| equisetum arvense | 28470,0 | 0,1 | caroube | 7752,0 | -0,7 |
| vinca minor | 28120,0 | 0,1 | boldo fragrans | 7821,0 | -0,7 |
| mentha piperita | 27120,0 | 0,1 | rhamnus purshiana | 7237,0 | -0,8 |
| sarsaparilla | 27100,0 | 0,1 | rosa canina | 7330,0 | -0,8 |
| spiruline nebl | 27212,0 | 0,1 | orthosiphon stamineus | 7040,0 | -0,8 |
| humulus lupulus | 26470,0 | 0,1 | harpagophytum proc | 5785,0 | -0,8 |
| betula alba | 24910,0 | 0,0 | gentiana lutea | 5587,0 | -0,8 |
| arctium lappa | 24720,0 | 0,0 | hypericum perforatum | 4897,0 | -0,9 |
| crataegus oxya | 24810,0 | 0,0 | melissa officinalis | 3950,0 | -0,9 |
| thymus vulgaris | 24550,0 | 0,0 | valeriana officinalis | 3467,0 | -0,9 |
| millefolium | 23440,0 | -0,1 | eugenia caryophyllata | 3255,0 | -0,9 |
| alchemilla vulgaris | 23200,0 | -0,1 | rhamnus frangula | 3334,0 | -0,9 |
| glycyrrhiza glabra | 22950,0 | -0,1 | ginseng | 3062,0 | -0,9 |
| senna | 21990,0 | -0,1 | chrysanthemum part | 2802,0 | -0,9 |
| carica papaya | 21620,0 | -0,2 | vaccinium myrtillus | 2114,0 | -1,0 |
| spiraea ulmaria | 21660,0 | -0,2 | china officinalis | 1296,0 | -1,0 |
| fucus vesiculosus | 19870,0 | -0,2 | viola tricolor | 1033,0 | -1,0 |
| agnus castus | 19780,0 | -0,2 | piper methysticum | 704,6 | -1,0 |
| beta vulgaris | 19170,0 | -0,3 | carduus marianus | 256,7 | -1,0 |
| erigeron canadensis | 18840,0 | -0,3 | althea officinalis | 80,7 | -1,1 |

# Selenium

| | | | | | |
|---|---|---|---|---|---|
| acorus calamus | 2,1 | 0,3 | hyoscyamus niger | 0,0 | -0,4 |
| aesculus hippocastanum | 0,0 | -0,4 | hypericum perforatum | 0,0 | -0,4 |
| agnus castus | 0,0 | -0,4 | juniperus communis | 0,0 | -0,4 |
| alchemilla vulgaris | 0,0 | -0,4 | lespedeza capitata | 0,0 | -0,4 |
| allium sativum | 0,0 | -0,4 | lithospermum | 0,0 | -0,4 |
| althea officinalis | 0,0 | -0,4 | lotus corniculatus | 0,0 | -0,4 |
| ananassa comosus | 0,0 | -0,4 | marrubium vulgare | 0,0 | -0,4 |
| angelica archangelica | 0,0 | -0,4 | medicago sativa | 0,0 | -0,4 |
| arctium lappa | 0,0 | -0,4 | melilotus officinalis | 0,0 | -0,4 |
| ballota foetida | 0,0 | -0,4 | melissa officinalis | 0,0 | -0,4 |
| berberis vulgaris | 2,4 | 0,4 | mentha piperita | 0,0 | -0,4 |
| beta vulgaris | 9,1 | 2,4 | millefolium | 0,0 | -0,4 |
| betula alba | 8,2 | 2,1 | nasturtium officinalis | 0,0 | -0,4 |
| boldo fragrans | 3,5 | 0,7 | olea europaea | 0,0 | -0,4 |
| carduus marianus | 2,3 | 0,3 | orthosiphon stamineus | 0,0 | -0,4 |
| carica papaya | 0,0 | -0,4 | passiflora incarnata | 0,0 | -0,4 |
| caroube | 0,0 | -0,4 | phaseolus vulgaris | 13,3 | 3,7 |
| carragheen | 0,0 | -0,4 | pinus maritima | 4,7 | 1,1 |
| carum carvi | 0,0 | -0,4 | piper methysticum | 0,0 | -0,4 |
| chamomilla | 9,5 | 2,5 | plantago lanceolata | 0,0 | -0,4 |
| china officinalis | 0,0 | -0,4 | raphanus sativus niger | 0,0 | -0,4 |
| chrysanthellum amer | 0,0 | -0,4 | rhamnus frangula | 7,2 | 1,8 |
| chrysanthemum part | 0,0 | -0,4 | rhamnus purshiana | 0,0 | -0,4 |
| crataegus oxyacantha | 0,0 | -0,4 | rheum officinale | 0,0 | -0,4 |
| cupressus semp | 0,0 | -0,4 | ribes nigrum folia | 0,0 | -0,4 |
| curcuma | 0,0 | -0,4 | ribes nigrum fructus | 0,0 | -0,4 |
| cynara scolymus | 0,0 | -0,4 | rosa canina | 0,0 | -0,4 |
| echinacea angustifolia | 0,0 | -0,4 | rosmarinus officinalis | 5,1 | 1,2 |
| echinacea purpurea | 10,5 | 2,8 | ruscus aculeatus | 0,0 | -0,4 |
| eleutherococcus | 0,0 | -0,4 | salix alba | 0,0 | -0,4 |
| equisetum arvense | 0,0 | -0,4 | salvia officinalis | 0,0 | -0,4 |
| erigeron canadensis | 0,0 | -0,4 | sarsaparilla | 0,0 | -0,4 |
| eschscholtzia cali | 0,0 | -0,4 | senna | 0,0 | -0,4 |
| eucalyptus | 0,0 | -0,4 | solidago virga aurea | 0,0 | -0,4 |
| eugenia caryophyllata | 0,0 | -0,4 | spiraea ulmaria | 5,6 | 1,3 |
| fenugrec | 0,0 | -0,4 | spiruline nebl | 0,0 | -0,4 |
| foeniculum vulgare | 8,8 | 2,3 | spiruline poud | 0,0 | -0,4 |
| fraxinus excelior | 0,0 | -0,4 | taraxacum | 0,0 | -0,4 |
| fucus vesiculosus | 0,0 | -0,4 | thymus vulgaris | 6,4 | 1,6 |
| fumaria officinalis | 0,0 | -0,4 | tilia alburnum | 0,0 | -0,4 |
| gentiana lutea | 0,0 | -0,4 | tilia cordata | 0,0 | -0,4 |
| ginkgo biloba | 0,0 | -0,4 | urtica dioica | 0,0 | -0,4 |
| ginseng | 0,0 | -0,4 | uva ursi | 1,8 | 0,2 |
| glycyrrhiza glabra | 0,0 | -0,4 | vaccinium myrtillus | 0,0 | -0,4 |
| hamamelis virginia | 0,0 | -0,4 | valeriana officinalis | 0,0 | -0,4 |
| harpagophytum proc | 0,0 | -0,4 | vinca minor | 0,0 | -0,4 |
| hibiscus sabdariffa | 0,0 | -0,4 | viola tricolor | 0,0 | -0,4 |
| hieracium pilosella | 0,0 | -0,4 | viscum album | 0,0 | -0,4 |
| humulus lupulus | 6,5 | 1,6 | vitis vinifera | 19,4 | 5,5 |
| hydrocotyle asiatica | 0,0 | -0,4 | zingiber officinalis | 0,0 | -0,4 |

# Selenium

| | | | | | |
|---|---|---|---|---|---|
| vitis vinifera | 19,4 | 5,5 | fumaria officinalis | 0,0 | -0,4 |
| phaseolus vulgaris | 13,3 | 3,7 | gentiana lutea | 0,0 | -0,4 |
| echinacea purpurea | 10,5 | 2,8 | ginkgo biloba | 0,0 | -0,4 |
| chamomilla | 9,5 | 2,5 | glycyrrhiza glabra | 0,0 | -0,4 |
| beta vulgaris | 9,1 | 2,4 | hibiscus sabdariffa | 0,0 | -0,4 |
| foeniculum vulgare | 8,8 | 2,3 | hamamelis virginia | 0,0 | -0,4 |
| betula alba | 8,2 | 2,1 | harpagophytum proc | 0,0 | -0,4 |
| rhamnus frangula | 7,2 | 1,8 | hieracium pilosella | 0,0 | -0,4 |
| humulus lupulus | 6,5 | 1,6 | hyoscyamus niger | 0,0 | -0,4 |
| thymus vulgaris | 6,4 | 1,6 | hypericum perforatum | 0,0 | -0,4 |
| spiraea ulmaria | 5,6 | 1,3 | juniperus communis | 0,0 | -0,4 |
| rosmarinus officinalis | 5,1 | 1,2 | lespedeza capitata | 0,0 | -0,4 |
| pinus maritima | 4,7 | 1,1 | lithospermum | 0,0 | -0,4 |
| boldo fragrans | 3,5 | 0,7 | lotus corniculatus | 0,0 | -0,4 |
| berberis vulgaris | 2,4 | 0,4 | marrubium vulgare | 0,0 | -0,4 |
| carduus marianus | 2,3 | 0,3 | medicago sativa | 0,0 | -0,4 |
| acorus calamus | 2,1 | 0,3 | melilotus officinalis | 0,0 | -0,4 |
| uva ursi | 1,8 | 0,2 | carica papaya | 0,0 | -0,4 |
| salvia officinalis | 0,0 | -0,4 | melissa officinalis | 0,0 | -0,4 |
| millefolium | 0,0 | -0,4 | mentha piperita | 0,0 | -0,4 |
| aesculus hippocastanum | 0,0 | -0,4 | nasturtium officinalis | 0,0 | -0,4 |
| alchemilla vulgaris | 0,0 | -0,4 | orthosiphon stamineus | 0,0 | -0,4 |
| allium sativum | 0,0 | -0,4 | ginseng | 0,0 | -0,4 |
| althea officinalis | 0,0 | -0,4 | passiflora incarnata | 0,0 | -0,4 |
| ananassa comosus | 0,0 | -0,4 | piper methysticum | 0,0 | -0,4 |
| angelica archangelica | 0,0 | -0,4 | plantago lanceolata | 0,0 | -0,4 |
| arctium lappa | 0,0 | -0,4 | china officinalis | 0,0 | -0,4 |
| ballota foetida | 0,0 | -0,4 | raphanus sativus niger | 0,0 | -0,4 |
| caroube | 0,0 | -0,4 | rhamnus purshiana | 0,0 | -0,4 |
| carragheen | 0,0 | -0,4 | rheum officinale | 0,0 | -0,4 |
| carum carvi | 0,0 | -0,4 | ribes nigrum folia | 0,0 | -0,4 |
| senna | 0,0 | -0,4 | ribes nigrum fructus | 0,0 | -0,4 |
| hydrocotyle asiatica | 0,0 | -0,4 | rosa canina | 0,0 | -0,4 |
| chrysanthellum amer | 0,0 | -0,4 | ruscus aculeatus | 0,0 | -0,4 |
| chrysanthemum part | 0,0 | -0,4 | salix alba | 0,0 | -0,4 |
| crataegus oxyacantha | 0,0 | -0,4 | sarsaparilla | 0,0 | -0,4 |
| cupressus semp | 0,0 | -0,4 | solidago virga aurea | 0,0 | -0,4 |
| curcuma | 0,0 | -0,4 | spiruline nebl | 0,0 | -0,4 |
| cynara scolymus | 0,0 | -0,4 | spiruline poud | 0,0 | -0,4 |
| echinacea angustifolia | 0,0 | -0,4 | taraxacum | 0,0 | -0,4 |
| olea europaea | 0,0 | -0,4 | tilia alburnum | 0,0 | -0,4 |
| eleutherococcus | 0,0 | -0,4 | tilia cordata | 0,0 | -0,4 |
| equisetum arvense | 0,0 | -0,4 | urtica dioica | 0,0 | -0,4 |
| erigeron canadensis | 0,0 | -0,4 | vaccinium myrtillus | 0,0 | -0,4 |
| eschscholtzia cali | 0,0 | -0,4 | valeriana officinalis | 0,0 | -0,4 |
| eucalyptus | 0,0 | -0,4 | vinca minor | 0,0 | -0,4 |
| eugenia caryophyllata | 0,0 | -0,4 | viola tricolor | 0,0 | -0,4 |
| fenugrec | 0,0 | -0,4 | viscum album | 0,0 | -0,4 |
| fraxinus excelior | 0,0 | -0,4 | agnus castus | 0,0 | -0,4 |
| fucus vesiculosus | 0,0 | -0,4 | zingiber officinalis | 0,0 | -0,4 |

# Sodium

| | | | | | | |
|---|---|---|---|---|---|---|
| acorus calamus | 6036,0 | 1,2 | hyoscyamus niger | 2800,0 | 0,2 |
| aesculus hipp | 491,5 | -0,5 | hypericum perforatum | 1614,0 | -0,1 |
| agnus castus | 788,4 | -0,4 | juniperus communis | 696,7 | -0,4 |
| alchemilla vulgaris | 94,0 | -0,6 | lespedeza capitata | 784,9 | -0,4 |
| allium sativum | 1198,0 | -0,2 | lithospermum | 936,6 | -0,3 |
| althea officinalis | 155,1 | -0,6 | lotus corniculatus | 1289,0 | -0,2 |
| ananassa comosus | 16,3 | -0,6 | marrubium vulgare | 1438,0 | -0,2 |
| angelica archangelica | 260,1 | -0,5 | medicago sativa | 1269,0 | -0,2 |
| arctium lappa | 1511,0 | -0,2 | melilotus officinalis | 823,3 | -0,4 |
| ballota foetida | 837,0 | -0,4 | melissa officinalis | 682,0 | -0,4 |
| berberis vulgaris | 1723,0 | -0,1 | mentha piperita | 2247,0 | 0,1 |
| beta vulgaris | 2049,0 | 0,0 | millefolium | 1321,0 | -0,2 |
| betula alba | 1226,0 | -0,2 | nasturtium officinalis | 7629,0 | 1,7 |
| boldo fragrans | 163,5 | -0,6 | olea europaea | 673,3 | -0,4 |
| carduus marianus | 1072,0 | -0,3 | orthosiphon stamineus | 344,1 | -0,5 |
| carica papaya | 1843,0 | -0,1 | passiflora incarnata | 1040,0 | -0,3 |
| caroube | 91,8 | -0,6 | phaseolus vulgaris | 454,2 | -0,5 |
| carragheen | 21350,0 | 5,7 | pinus maritima | 778,9 | -0,4 |
| carum carvi | 665,7 | -0,4 | piper methysticum | 212,6 | -0,5 |
| chamomilla | 4400,0 | 0,7 | plantago lanceolata | 1944,0 | 0,0 |
| china officinalis | 1638,0 | -0,1 | raphanus sativus niger | 1888,0 | 0,0 |
| chrysanthellum amer | 267,4 | -0,5 | rhamnus frangula | 825,6 | -0,4 |
| chrysanthemum part | 467,0 | -0,5 | rhamnus purshiana | 266,9 | -0,5 |
| crataegus oxyacantha | 2008,0 | 0,0 | rheum officinale | 2581,0 | 0,2 |
| cupressus semp | 2890,0 | 0,3 | ribes nigrum folia | 80,7 | -0,6 |
| curcuma | 368,8 | -0,5 | ribes nigrum fructus | 242,9 | -0,5 |
| cynara scolymus | 3338,0 | 0,4 | rosa canina | 580,0 | -0,4 |
| echinacea angustifolia | 1995,0 | 0,0 | rosmarinus officinalis | 1861,0 | 0,0 |
| echinacea purpurea | 2412,0 | 0,1 | ruscus aculeatus | 732,1 | -0,4 |
| eleutherococcus | 172,2 | -0,5 | salix alba | 791,8 | -0,4 |
| equisetum arvense | 1236,0 | -0,2 | salvia officinalis | 625,5 | -0,4 |
| erigeron canadensis | 109,0 | -0,6 | sarsaparilla | 1106,0 | -0,3 |
| eschscholtzia cali | 1118,0 | -0,3 | senna | 1544,0 | -0,1 |
| eucalyptus | 2433,0 | 0,1 | solidago virga aurea | 335,9 | -0,5 |
| eugenia caryophyllata | 2253,0 | 0,1 | spiraea ulmaria | 908,4 | -0,3 |
| fenugrec | 258,9 | -0,5 | spiruline nebl | 3305,0 | 0,4 |
| foeniculum vulgare | 5136,0 | 0,9 | spiruline poud | 12390,0 | 3,1 |
| fraxinus excelior | 2287,0 | 0,1 | taraxacum | 8746,0 | 2,0 |
| fucus vesiculosus | 21220,0 | 5,7 | thymus vulgaris | 2054,0 | 0,0 |
| fumaria officinalis | 570,3 | -0,4 | tilia alburnum | 631,5 | -0,4 |
| gentiana lutea | 2557,0 | 0,2 | tilia cordata | 979,4 | -0,3 |
| ginkgo biloba | 1012,0 | -0,3 | urtica dioica | 1755,0 | -0,1 |
| ginseng | 2676,0 | 0,2 | uva ursi | 581,3 | -0,4 |
| glycyrrhiza glabra | 6973,0 | 1,5 | vaccinium myrtillus | 332,8 | -0,5 |
| hamamelis virginia | 401,6 | -0,5 | valeriana officinalis | 1007,0 | -0,3 |
| harpagophytum proc | 1227,0 | -0,2 | vinca minor | 1206,0 | -0,2 |
| hibiscus sabdariffa | 497,8 | -0,5 | viola tricolor | 860,5 | -0,3 |
| hieracium pilosella | 1144,0 | -0,3 | viscum album | 479,9 | -0,5 |
| humulus lupulus | 125,5 | -0,6 | vitis vinifera | 846,0 | -0,3 |
| hydrocotyle asiatica | 5774,0 | 1,1 | zingiber officinalis | 1436,0 | -0,2 |

# Sodium

| | | | | | |
|---|---:|---:|---|---:|---:|
| carragheen | 21350,0 | 5,7 | ginkgo biloba | 1012,0 | -0,3 |
| fucus vesiculosus | 21220,0 | 5,7 | passiflora incarnata | 1040,0 | -0,3 |
| spiruline poud | 12390,0 | 3,1 | tilia cordata | 979,4 | -0,3 |
| taraxacum | 8746,0 | 2,0 | valeriana officinalis | 1007,0 | -0,3 |
| nasturtium officinalis | 7629,0 | 1,7 | lithospermum | 936,6 | -0,3 |
| glycyrrhiza glabra | 6973,0 | 1,5 | spiraea ulmaria | 908,4 | -0,3 |
| acorus calamus | 6036,0 | 1,2 | viola tricolor | 860,5 | -0,3 |
| hydrocotyle asiatica | 5774,0 | 1,1 | vitis vinifera | 846,0 | -0,3 |
| foeniculum vulgare | 5136,0 | 0,9 | ballota foetida | 837,0 | -0,4 |
| chamomilla | 4400,0 | 0,7 | melilotus officinalis | 823,3 | -0,4 |
| cynara scolymus | 3338,0 | 0,4 | rhamnus frangula | 825,6 | -0,4 |
| spiruline nebl | 3305,0 | 0,4 | lespedeza capitata | 784,9 | -0,4 |
| cupressus semp | 2890,0 | 0,3 | pinus maritima | 778,9 | -0,4 |
| hyoscyamus niger | 2800,0 | 0,2 | salix alba | 791,8 | -0,4 |
| ginseng | 2676,0 | 0,2 | agnus castus | 788,4 | -0,4 |
| rheum officinale | 2581,0 | 0,2 | ruscus aculeatus | 732,1 | -0,4 |
| gentiana lutea | 2557,0 | 0,2 | olea europaea | 673,3 | -0,4 |
| eucalyptus | 2433,0 | 0,1 | juniperus communis | 696,7 | -0,4 |
| echinacea purpurea | 2412,0 | 0,1 | melissa officinalis | 682,0 | -0,4 |
| fraxinus excelior | 2287,0 | 0,1 | carum carvi | 665,7 | -0,4 |
| eugenia caryophyllata | 2253,0 | 0,1 | salvia officinalis | 625,5 | -0,4 |
| mentha piperita | 2247,0 | 0,1 | tilia alburnum | 631,5 | -0,4 |
| beta vulgaris | 2049,0 | 0,0 | uva ursi | 581,3 | -0,4 |
| thymus vulgaris | 2054,0 | 0,0 | fumaria officinalis | 570,3 | -0,4 |
| crataegus oxyacantha | 2008,0 | 0,0 | rosa canina | 580,0 | -0,4 |
| echinacea angustifolia | 1995,0 | 0,0 | aesculus hippocastanum | 491,5 | -0,5 |
| plantago lanceolata | 1944,0 | 0,0 | hibiscus sabdariffa | 497,8 | -0,5 |
| raphanus sativus niger | 1888,0 | 0,0 | viscum album | 479,9 | -0,5 |
| rosmarinus officinalis | 1861,0 | 0,0 | chrysanthemum part | 467,0 | -0,5 |
| carica papaya | 1843,0 | -0,1 | phaseolus vulgaris | 454,2 | -0,5 |
| urtica dioica | 1755,0 | -0,1 | hamamelis virginia | 401,6 | -0,5 |
| berberis vulgaris | 1723,0 | -0,1 | curcuma | 368,8 | -0,5 |
| china officinalis | 1638,0 | -0,1 | orthosiphon stamineus | 344,1 | -0,5 |
| hypericum perforatum | 1614,0 | -0,1 | solidago virga aurea | 335,9 | -0,5 |
| senna | 1544,0 | -0,1 | vaccinium myrtillus | 332,8 | -0,5 |
| arctium lappa | 1511,0 | -0,2 | chrysanthellum amer | 267,4 | -0,5 |
| marrubium vulgare | 1438,0 | -0,2 | rhamnus purshiana | 266,9 | -0,5 |
| zingiber officinalis | 1436,0 | -0,2 | angelica archangelica | 260,1 | -0,5 |
| millefolium | 1321,0 | -0,2 | fenugrec | 258,9 | -0,5 |
| lotus corniculatus | 1289,0 | -0,2 | ribes nigrum fructus | 242,9 | -0,5 |
| medicago sativa | 1269,0 | -0,2 | piper methysticum | 212,6 | -0,5 |
| betula alba | 1226,0 | -0,2 | eleutherococcus | 172,2 | -0,5 |
| equisetum arvense | 1236,0 | -0,2 | althea officinalis | 155,1 | -0,6 |
| harpagophytum proc | 1227,0 | -0,2 | boldo fragrans | 163,5 | -0,6 |
| allium sativum | 1198,0 | -0,2 | erigeron canadensis | 109,0 | -0,6 |
| vinca minor | 1206,0 | -0,2 | humulus lupulus | 125,5 | -0,6 |
| eschscholtzia cali | 1118,0 | -0,3 | alchemilla vulgaris | 94,0 | -0,6 |
| hieracium pilosella | 1144,0 | -0,3 | caroube | 91,8 | -0,6 |
| sarsaparilla | 1106,0 | -0,3 | ribes nigrum folia | 80,7 | -0,6 |
| carduus marianus | 1072,0 | -0,3 | ananassa comosus | 16,3 | -0,6 |

# Vanadium

| | | | | | | |
|---|---|---|---|---|---|---|
| acorus calamus | 0,3 | -0,3 | hyoscyamus niger | 0,2 | -0,4 |
| aesculus hippocastanum | 0,1 | -0,5 | hypericum perforatum | 0,0 | -0,7 |
| agnus castus | 0,5 | 0,0 | juniperus communis | 0,4 | -0,1 |
| alchemilla vulgaris | 0,0 | -0,7 | lespedeza capitata | 0,0 | -0,7 |
| allium sativum | 0,0 | -0,7 | lithospermum | 0,8 | 0,5 |
| althea officinalis | 1,3 | 1,2 | lotus corniculatus | 0,7 | 0,3 |
| ananassa comosus | 0,8 | 0,5 | marrubium vulgare | 0,0 | -0,7 |
| angelica archangelica | 0,0 | -0,7 | medicago sativa | 1,5 | 1,5 |
| arctium lappa | 0,0 | -0,7 | melilotus officinalis | 1,3 | 1,2 |
| ballota foetida | 0,8 | 0,5 | melissa officinalis | 0,7 | 0,3 |
| berberis vulgaris | 0,6 | 0,2 | mentha piperita | 2,9 | 3,6 |
| beta vulgaris | 0,1 | -0,6 | millefolium | 0,0 | -0,7 |
| betula alba | 0,0 | -0,7 | nasturtium officinalis | 1,9 | 2,1 |
| boldo fragrans | 0,0 | -0,7 | olea europaea | 0,2 | -0,4 |
| carduus marianus | 0,0 | -0,7 | orthosiphon stamineus | 0,0 | -0,7 |
| carica papaya | 0,5 | 0,0 | passiflora incarnata | 0,4 | -0,1 |
| caroube | 2,8 | 3,4 | phaseolus vulgaris | 0,0 | -0,7 |
| carragheen | 1,3 | 1,2 | pinus maritima | 0,0 | -0,7 |
| carum carvi | 0,3 | -0,2 | piper methysticum | 0,0 | -0,7 |
| chamomilla | 0,0 | -0,7 | plantago lanceolata | 1,0 | 0,8 |
| china officinalis | 0,0 | -0,7 | raphanus sativus niger | 0,8 | 0,5 |
| chrysanthellum amer | 0,4 | -0,1 | rhamnus frangula | 0,0 | -0,7 |
| chrysanthemum part | 0,0 | -0,7 | rhamnus purshiana | 0,0 | -0,7 |
| crataegus oxyacantha | 0,0 | -0,7 | rheum officinale | 0,0 | -0,7 |
| cupressus semp | 0,6 | 0,2 | ribes nigrum folia | 0,9 | 0,6 |
| curcuma | 0,0 | -0,7 | ribes nigrum fructus | 0,0 | -0,7 |
| cynara scolymus | 0,0 | -0,7 | rosa canina | 0,0 | -0,7 |
| echinacea angustifolia | 0,0 | -0,7 | rosmarinus officinalis | 0,8 | 0,5 |
| echinacea purpurea | 0,0 | -0,7 | ruscus aculeatus | 0,0 | -0,7 |
| eleutherococcus | 0,0 | -0,7 | salix alba | 2,6 | 3,1 |
| equisetum arvense | 0,0 | -0,7 | salvia officinalis | 1,0 | 0,8 |
| erigeron canadensis | 0,0 | -0,7 | sarsaparilla | 0,6 | 0,2 |
| eschscholtzia cali | 0,5 | 0,0 | senna | 0,0 | -0,7 |
| eucalyptus | 0,3 | -0,3 | solidago virga aurea | 1,0 | 0,8 |
| eugenia caryophyllata | 0,0 | -0,7 | spiraea ulmaria | 0,0 | -0,7 |
| fenugrec | 0,0 | -0,7 | spiruline nebl | 0,6 | 0,2 |
| foeniculum vulgare | 0,0 | -0,7 | spiruline poud | 0,0 | -0,7 |
| fraxinus excelior | 1,7 | 1,8 | taraxacum | 1,3 | 1,2 |
| fucus vesiculosus | 0,0 | -0,7 | thymus vulgaris | 2,4 | 2,8 |
| fumaria officinalis | 0,2 | -0,4 | tilia alburnum | 1,0 | 0,8 |
| gentiana lutea | 0,0 | -0,7 | tilia cordata | 0,2 | -0,4 |
| ginkgo biloba | 0,1 | -0,6 | urtica dioica | 1,5 | 1,5 |
| ginseng | 0,7 | 0,3 | uva ursi | 0,0 | -0,7 |
| glycyrrhiza glabra | 1,3 | 1,2 | vaccinium myrtillus | 0,0 | -0,7 |
| hamamelis virginia | 0,6 | 0,2 | valeriana officinalis | 0,0 | -0,7 |
| harpagophytum proc | 0,0 | -0,7 | vinca minor | 0,3 | -0,3 |
| hibiscus sabdariffa | 2,1 | 2,4 | viola tricolor | 1,0 | 0,8 |
| hieracium pilosella | 1,3 | 1,2 | viscum album | 0,0 | -0,7 |
| humulus lupulus | 0,0 | -0,7 | vitis vinifera | 0,8 | 0,5 |
| hydrocotyle asiatica | 0,0 | -0,7 | zingiber officinalis | 0,1 | -0,6 |

# Vanadium

| | | | | | |
|---|---|---|---|---|---|
| mentha piperita | 2,9 | 3,6 | beta vulgaris | 0,1 | -0,6 |
| caroube | 2,8 | 3,4 | ginkgo biloba | 0,1 | -0,6 |
| salix alba | 2,6 | 3,1 | zingiber officinalis | 0,1 | -0,6 |
| thymus vulgaris | 2,4 | 2,8 | millefolium | 0,0 | -0,7 |
| hibiscus sabdariffa | 2,1 | 2,4 | alchemilla vulgaris | 0,0 | -0,7 |
| nasturtium officinalis | 1,9 | 2,1 | allium sativum | 0,0 | -0,7 |
| fraxinus excelior | 1,7 | 1,8 | angelica archangelica | 0,0 | -0,7 |
| medicago sativa | 1,5 | 1,5 | arctium lappa | 0,0 | -0,7 |
| urtica dioica | 1,5 | 1,5 | uva ursi | 0,0 | -0,7 |
| althea officinalis | 1,3 | 1,2 | betula alba | 0,0 | -0,7 |
| carragheen | 1,3 | 1,2 | carduus marianus | 0,0 | -0,7 |
| glycyrrhiza glabra | 1,3 | 1,2 | senna | 0,0 | -0,7 |
| hieracium pilosella | 1,3 | 1,2 | hydrocotyle asiatica | 0,0 | -0,7 |
| melilotus officinalis | 1,3 | 1,2 | chrysanthemum part | 0,0 | -0,7 |
| taraxacum | 1,3 | 1,2 | crataegus oxyacantha | 0,0 | -0,7 |
| plantago lanceolata | 1,0 | 0,8 | curcuma | 0,0 | -0,7 |
| salvia officinalis | 1,0 | 0,8 | cynara scolymus | 0,0 | -0,7 |
| solidago virga aurea | 1,0 | 0,8 | echinacea angustifolia | 0,0 | -0,7 |
| tilia alburnum | 1,0 | 0,8 | echinacea purpurea | 0,0 | -0,7 |
| viola tricolor | 1,0 | 0,8 | eleutherococcus | 0,0 | -0,7 |
| ribes nigrum folia | 0,9 | 0,6 | equisetum arvense | 0,0 | -0,7 |
| ananassa comosus | 0,8 | 0,5 | erigeron canadensis | 0,0 | -0,7 |
| ballota foetida | 0,8 | 0,5 | eugenia caryophyllata | 0,0 | -0,7 |
| lithospermum | 0,8 | 0,5 | fenugrec | 0,0 | -0,7 |
| raphanus sativus niger | 0,8 | 0,5 | foeniculum vulgare | 0,0 | -0,7 |
| rosmarinus officinalis | 0,8 | 0,5 | fucus vesiculosus | 0,0 | -0,7 |
| vitis vinifera | 0,8 | 0,5 | gentiana lutea | 0,0 | -0,7 |
| lotus corniculatus | 0,7 | 0,3 | boldo fragrans | 0,0 | -0,7 |
| melissa officinalis | 0,7 | 0,3 | harpagophytum proc | 0,0 | -0,7 |
| ginseng | 0,7 | 0,3 | humulus lupulus | 0,0 | -0,7 |
| berberis vulgaris | 0,6 | 0,2 | hypericum perforatum | 0,0 | -0,7 |
| cupressus semp | 0,6 | 0,2 | lespedeza capitata | 0,0 | -0,7 |
| hamamelis virginia | 0,6 | 0,2 | marrubium vulgare | 0,0 | -0,7 |
| sarsaparilla | 0,6 | 0,2 | chamomilla | 0,0 | -0,7 |
| spiruline nebl | 0,6 | 0,2 | orthosiphon stamineus | 0,0 | -0,7 |
| eschscholtzia cali | 0,5 | 0,0 | phaseolus vulgaris | 0,0 | -0,7 |
| carica papaya | 0,5 | 0,0 | pinus maritima | 0,0 | -0,7 |
| agnus castus | 0,5 | 0,0 | piper methysticum | 0,0 | -0,7 |
| chrysanthellum amer | 0,4 | -0,1 | china officinalis | 0,0 | -0,7 |
| juniperus communis | 0,4 | -0,1 | rhamnus frangula | 0,0 | -0,7 |
| passiflora incarnata | 0,4 | -0,1 | rhamnus purshiana | 0,0 | -0,7 |
| carum carvi | 0,3 | -0,2 | rheum officinale | 0,0 | -0,7 |
| acorus calamus | 0,3 | -0,3 | ribes nigrum fructus | 0,0 | -0,7 |
| eucalyptus | 0,3 | -0,3 | rosa canina | 0,0 | -0,7 |
| vinca minor | 0,3 | -0,3 | ruscus aculeatus | 0,0 | -0,7 |
| olea europaea | 0,2 | -0,4 | spiraea ulmaria | 0,0 | -0,7 |
| fumaria officinalis | 0,2 | -0,4 | spiruline poud | 0,0 | -0,7 |
| hyoscyamus niger | 0,2 | -0,4 | vaccinium myrtillus | 0,0 | -0,7 |
| tilia cordata | 0,2 | -0,4 | valeriana officinalis | 0,0 | -0,7 |
| aesculus hippocastanum | 0,1 | -0,5 | viscum album | 0,0 | -0,7 |

# Zinc

| | | | | | |
|---|---|---|---|---|---|
| acorus calamus | 4,7 | -0,7 | hyoscyamus niger | 34,4 | 1,0 |
| aesculus hippocastanum | 5,8 | -0,6 | hypericum perforatum | 6,7 | -0,6 |
| agnus castus | 5,3 | -0,6 | juniperus communis | 4,8 | -0,7 |
| alchemilla vulgaris | 29,2 | 0,7 | lespedeza capitata | 20,1 | 0,2 |
| allium sativum | 22,0 | 0,3 | lithospermum | 25,4 | 0,5 |
| althea officinalis | 6,8 | -0,6 | lotus corniculatus | 46,0 | 1,7 |
| ananassa comosus | 7,4 | -0,5 | marrubium vulgare | 13,5 | -0,2 |
| angelica archangelica | 3,5 | -0,7 | medicago sativa | 9,1 | -0,4 |
| arctium lappa | 6,5 | -0,6 | melilotus officinalis | 23,2 | 0,4 |
| ballota foetida | 17,1 | 0,0 | melissa officinalis | 2,9 | -0,8 |
| berberis vulgaris | 21,9 | 0,3 | mentha piperita | 5,3 | -0,6 |
| beta vulgaris | 27,7 | 0,6 | millefolium | 7,2 | -0,5 |
| betula alba | 51,4 | 2,0 | nasturtium officinalis | 34,7 | 1,0 |
| boldo fragrans | 2,4 | -0,8 | olea europaea | 3,7 | -0,7 |
| carduus marianus | 3,8 | -0,7 | orthosiphon stamineus | 20,9 | 0,3 |
| carica papaya | 3,8 | -0,7 | passiflora incarnata | 49,9 | 1,9 |
| caroube | 8,4 | -0,5 | phaseolus vulgaris | 50,4 | 1,9 |
| carragheen | 46,4 | 1,7 | pinus maritima | 97,2 | 4,6 |
| carum carvi | 4,6 | -0,7 | piper methysticum | 1,5 | -0,9 |
| chamomilla | 8,4 | -0,5 | plantago lanceolata | 5,5 | -0,6 |
| china officinalis | 7,2 | -0,5 | raphanus sativus niger | 16,5 | 0,0 |
| chrysanthellum amer | 14,9 | -0,1 | rhamnus frangula | 6,5 | -0,6 |
| chrysanthemum part | 8,0 | -0,5 | rhamnus purshiana | 3,9 | -0,7 |
| crataegus oxyacantha | 24,2 | 0,4 | rheum officinale | 3,0 | -0,8 |
| cupressus semp | 13,7 | -0,2 | ribes nigrum folia | 6,7 | -0,6 |
| curcuma | 16,3 | 0,0 | ribes nigrum fructus | 5,0 | -0,7 |
| cynara scolymus | 4,2 | -0,7 | rosa canina | 2,7 | -0,8 |
| echinacea angustifolia | 13,7 | -0,2 | rosmarinus officinalis | 7,0 | -0,5 |
| echinacea purpurea | 13,1 | -0,2 | ruscus aculeatus | 3,3 | -0,8 |
| eleutherococcus | 4,3 | -0,7 | salix alba | 59,8 | 2,5 |
| equisetum arvense | 4,4 | -0,7 | salvia officinalis | 6,6 | -0,6 |
| erigeron canadensis | 9,1 | -0,4 | sarsaparilla | 9,9 | -0,4 |
| eschscholtzia cali | 84,3 | 3,9 | senna | 41,6 | 1,4 |
| eucalyptus | 6,1 | -0,6 | solidago virga aurea | 25,7 | 0,5 |
| eugenia caryophyllata | 2,4 | -0,8 | spiraea ulmaria | 14,9 | -0,1 |
| fenugrec | 5,1 | -0,7 | spiruline nebl | 5,8 | -0,6 |
| foeniculum vulgare | 17,0 | 0,0 | spiruline poud | 16,6 | 0,0 |
| fraxinus excelior | 8,4 | -0,5 | taraxacum | 4,4 | -0,7 |
| fucus vesiculosus | 3,8 | -0,7 | thymus vulgaris | 14,2 | -0,1 |
| fumaria officinalis | 24,4 | 0,5 | tilia alburnum | 5,1 | -0,7 |
| gentiana lutea | 9,7 | -0,4 | tilia cordata | 8,9 | -0,4 |
| ginkgo biloba | 7,5 | -0,5 | urtica dioica | 38,6 | 1,3 |
| ginseng | 5,4 | -0,6 | uva ursi | 9,5 | -0,4 |
| glycyrrhiza glabra | 10,9 | -0,3 | vaccinium myrtillus | 5,5 | -0,6 |
| hamamelis virginia | 5,3 | -0,6 | valeriana officinalis | 1,5 | -0,9 |
| harpagophytum proc | 3,4 | -0,8 | vinca minor | 17,8 | 0,1 |
| hibiscus sabdariffa | 31,6 | 0,9 | viola tricolor | 40,0 | 1,3 |
| hieracium pilosella | 31,5 | 0,9 | viscum album | 28,3 | 0,7 |
| humulus lupulus | 0,7 | -0,9 | vitis vinifera | 32,8 | 0,9 |
| hydrocotyle asiatica | 17,9 | 0,1 | zingiber officinalis | 50,2 | 1,9 |

# Zinc

| | | | | |
|---|---|---|---|---|
| pinus maritima | 97,2 | 4,6 | caroube | 8,4 | -0,5 |
| eschscholtzia cali | 84,3 | 3,9 | fraxinus excelior | 8,4 | -0,5 |
| salix alba | 59,8 | 2,5 | chamomilla | 8,4 | -0,5 |
| betula alba | 51,4 | 2,0 | chrysanthemum part | 8,0 | -0,5 |
| phaseolus vulgaris | 50,4 | 1,9 | ginkgo biloba | 7,5 | -0,5 |
| zingiber officinalis | 50,2 | 1,9 | ananassa comosus | 7,4 | -0,5 |
| passiflora incarnata | 49,9 | 1,9 | millefolium | 7,2 | -0,5 |
| carragheen | 46,4 | 1,7 | china officinalis | 7,2 | -0,5 |
| lotus corniculatus | 46,0 | 1,7 | rosmarinus officinalis | 7,0 | -0,5 |
| senna | 41,6 | 1,4 | althea officinalis | 6,8 | -0,6 |
| viola tricolor | 40,0 | 1,3 | hypericum perforatum | 6,7 | -0,6 |
| urtica dioica | 38,6 | 1,3 | ribes nigrum folia | 6,7 | -0,6 |
| nasturtium officinalis | 34,7 | 1,0 | arctium lappa | 6,5 | -0,6 |
| hyoscyamus niger | 34,4 | 1,0 | rhamnus frangula | 6,5 | -0,6 |
| vitis vinifera | 32,8 | 0,9 | salvia officinalis | 6,6 | -0,6 |
| hibiscus sabdariffa | 31,6 | 0,9 | eucalyptus | 6,1 | -0,6 |
| hieracium pilosella | 31,5 | 0,9 | aesculus hippocastanum | 5,8 | -0,6 |
| alchemilla vulgaris | 29,2 | 0,7 | spiruline nebl | 5,8 | -0,6 |
| viscum album | 28,3 | 0,7 | ginseng | 5,4 | -0,6 |
| beta vulgaris | 27,7 | 0,6 | plantago lanceolata | 5,5 | -0,6 |
| solidago virga aurea | 25,7 | 0,5 | vaccinium myrtillus | 5,5 | -0,6 |
| lithospermum | 25,4 | 0,5 | hamamelis virginia | 5,3 | -0,6 |
| fumaria officinalis | 24,4 | 0,5 | mentha piperita | 5,3 | -0,6 |
| crataegus oxyacantha | 24,2 | 0,4 | agnus castus | 5,3 | -0,6 |
| melilotus officinalis | 23,2 | 0,4 | fenugrec | 5,1 | -0,7 |
| allium sativum | 22,0 | 0,3 | tilia alburnum | 5,1 | -0,7 |
| berberis vulgaris | 21,9 | 0,3 | ribes nigrum fructus | 5,0 | -0,7 |
| orthosiphon stamineus | 20,9 | 0,3 | acorus calamus | 4,7 | -0,7 |
| lespedeza capitata | 20,1 | 0,2 | juniperus communis | 4,8 | -0,7 |
| hydrocotyle asiatica | 17,9 | 0,1 | carum carvi | 4,6 | -0,7 |
| vinca minor | 17,8 | 0,1 | equisetum arvense | 4,4 | -0,7 |
| ballota foetida | 17,1 | 0,0 | taraxacum | 4,4 | -0,7 |
| foeniculum vulgare | 17,0 | 0,0 | cynara scolymus | 4,2 | -0,7 |
| spiruline poud | 16,6 | 0,0 | eleutherococcus | 4,3 | -0,7 |
| raphanus sativus niger | 16,5 | 0,0 | carduus marianus | 3,8 | -0,7 |
| curcuma | 16,3 | 0,0 | fucus vesiculosus | 3,8 | -0,7 |
| chrysanthellum amer | 14,9 | -0,1 | carica papaya | 3,8 | -0,7 |
| spiraea ulmaria | 14,9 | -0,1 | rhamnus purshiana | 3,9 | -0,7 |
| thymus vulgaris | 14,2 | -0,1 | olea europaea | 3,7 | -0,7 |
| cupressus semp | 13,7 | -0,2 | angelica archangelica | 3,5 | -0,7 |
| echinacea angustifolia | 13,7 | -0,2 | harpagophytum proc | 3,4 | -0,8 |
| marrubium vulgare | 13,5 | -0,2 | ruscus aculeatus | 3,3 | -0,8 |
| echinacea purpurea | 13,1 | -0,2 | rheum officinale | 3,0 | -0,8 |
| glycyrrhiza glabra | 10,9 | -0,3 | melissa officinalis | 2,9 | -0,8 |
| sarsaparilla | 9,9 | -0,4 | rosa canina | 2,7 | -0,8 |
| gentiana lutea | 9,7 | -0,4 | eugenia caryophyllata | 2,4 | -0,8 |
| uva ursi | 9,5 | -0,4 | boldo fragrans | 2,4 | -0,8 |
| erigeron canadensis | 9,1 | -0,4 | piper methysticum | 1,5 | -0,9 |
| medicago sativa | 9,1 | -0,4 | valeriana officinalis | 1,5 | -0,9 |
| tilia cordata | 8,9 | -0,4 | humulus lupulus | 0,7 | -0,9 |